AF843546

Enzyme Nanocarriers

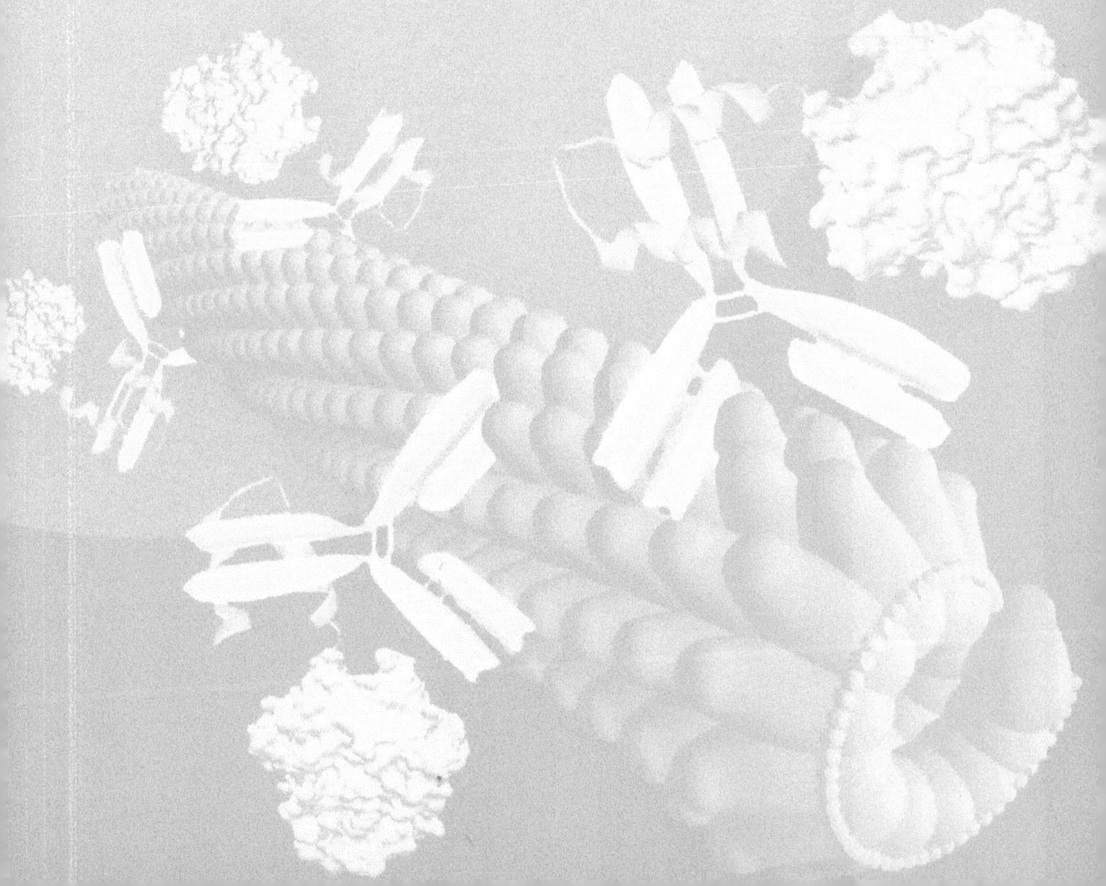

Enzyme Nanocarriers

edited by

Daniela Cardinale
Thierry Michon

Pan Stanford Publishing

Published by

Pan Stanford Publishing Pte. Ltd.
Penthouse Level, Suntec Tower 3
8 Temasek Boulevard
Singapore 038988

Email: editorial@panstanford.com
Web: www.panstanford.com

British Library Cataloguing-in-Publication Data
A catalogue record for this book is available from the British Library.

Enzyme Nanocarriers

Copyright © 2016 by Pan Stanford Publishing Pte. Ltd.
All rights reserved. This book, or parts thereof, may not be reproduced in any form or by any means, electronic or mechanical, including photocopying, recording or any information storage and retrieval system now known or to be invented, without written permission from the publisher.

For photocopying of material in this volume, please pay a copying fee through the Copyright Clearance Center, Inc., 222 Rosewood Drive, Danvers, MA 01923, USA. In this case permission to photocopy is not required from the publisher.

ISBN 978-981-4613-42-2 (Hardcover)
ISBN 978-981-4613-43-9 (eBook)

Cover graphic conception: Daniela Cardinale and Mark B. van Eldijik
Special thanks to www.cronodon.com

Printed in the USA

Contents

Preface ix
Introduction xi
Acknowledgment xix

1. Virus Diversity to Explore Various Kinds of Enzyme Nanocarriers **1**
Jane Besong-Ndika, Jocelyne Walter, and Kristiina Mäkinen
1.1 Introduction to Virus Diversity 1
1.2 Assembly and Packaging Strategies of Viruses 5
1.2.1 Particle Formation via Coassembly 6
1.2.1.1 Assembly of TMV 6
1.2.2 Particle Formation Via Preformed Procapsids 10
1.3 Virion Formation and Its Multiple Roles in Virus Infection 11
1.4 Production of VLPs 15
1.4.1 Helical Viruses 15
1.4.2 Spherical Viruses 16
1.4.3 VLP Expression Systems 19
1.5 Various Virion Structures and Properties for ENC Applications 20
1.5.1 Icosahedral Viruses 21
1.5.2 Helical Viruses 25
1.5.3 Enveloped Viruses 28
1.6 New Viral Species to Be Exploited as Viral ENCs 29

2. Strategies for Virus–Enzyme Coupling **45**
Daniela Cardinale, Noëlle Carette, and Thierry Michon
2.1 Covalent Attachment of Enzymes to Virus Nanoparticles 45
2.1.1 Protein Engineering for Targeting Specific Coupling Sites 50
2.1.1.1 Natural amino acids 50
2.1.1.2 Copper(I)-catalyzed azide–alkyne cycloaddition 59

2.1.1.3 Unnatural Amino Acids 61
2.2 Genetic Strategies for Direct Enzyme Fusion 64
2.2.1 Nonenveloped Viruses 66
2.2.1.1 Plant viruses 66
2.2.1.2 Bacteriophages 67
2.2.1.3 Adenoviruses 71
2.2.1.4 Rotaviruses 72
2.2.1.5 Parvoviruses 73
2.2.1.6 Core-like particles 73
2.2.2 Enveloped Viruses 75
2.2.2.1 Modification of viral proteins 76
2.2.2.2 Membrane anchoring 82
2.3 Coupling of Adaptors 86
2.3.1 Peptides from Phage Display: The "Nanoglue" Concept 86
2.3.2 Biotin–Streptavidin 87
2.3.3 Nucleic Acid–Guided Packaging 88
2.3.4 Coiled Coils 89
2.4 Immunostrategies 89
2.4.1 Bispecific Antibodies 89
2.4.2 Z33 90

3. Viruses as Model Nanoreactors to Study Enzyme Kinetics **105**

W. Frederik Rurup, Melissa S. T. Koay, and Jeroen J. L. M. Cornelissen

3.1 Introduction 107
3.2 Virus-Based Building Blocks for Nanoreactors 107
3.2.1 Plant-Based Viruses 107
3.2.2 Bacteriophages 110
3.3 Enzyme-Loading Strategies 111
3.3.1 In vitro Loading 111
3.3.1.1 Coiled-coil peptides 111
3.3.1.2 Charge-induced loading 112
3.3.2 In vivo Loading 113
3.3.2.1 DNA aptamers 113
3.3.2.2 Signalling proteins 113

3.3.3 Essential differences in loading strategies 114
3.4 Encapsulation Effects 114
3.4.1 CCMV 115
3.4.2 Bacteriophages MS2 and Qβ 115
3.4.3 Bacteriophage P22 116
3.5 Substrate Diffusion 117
3.6 Future Outlook 118

4. Nanoscale Compartmentalization Techniques in Cascade Catalysis **123**

Jorgen S. Willemsen, Floris P. J. T. Rutjes, and Jan C. M. van Hest

4.1 Cascade Reactions and Incompatibility Issues 123
4.2 Microdroplets 128
4.3 Compartmentalization in Polymeric Nanoparticles and Micelles 129
4.4 Templated Nanocapsules 132
4.5 Liposomes 137
4.6 Polymersomes 139
4.7 Viruses and Virus-Like Particles 147
4.8 Outlook 148

5. Nanocarriers for Therapeutic Enzymes **159**

Melissa D. Howard, Elizabeth D. Hood, and Vladimir R. Muzykantov

5.1 Introduction 159
5.2 Nanocarriers 160
5.2.1 Rationale for the Use of Nanocarriers 160
5.2.2 Types of Nanocarriers 161
5.3 Cargos 164
5.3.1 Fibrinolytics 164
5.3.2 Antioxidant Enzymes 169
5.3.3 Enzymes for Antitumor Therapy 172
5.3.4 Detoxifying Enzymes 174
5.3.5 Enzyme Replacement Therapy 175
5.3.5.1 Systemic ERT 175
5.3.5.2 Lysosomal storage disorders 177
5.4 Conclusion: Translational Outlook, Challenges, and Perspectives 180

6. DNA-Origamis as Protein Nanocarriers **193**
Juan Elezgaray, Jean-Pierre Aimé, and Jean-Michel Arbona
6.1 A Quick Primer to DNA Origami Formation 194
6.2 Boxes and Nanorobots 196
6.2.1 Nanorobots for Drug Delivery 197
6.3 Protein and Nucleic Acid Arrays 199
6.3.1 Arranging Nucleic Acid Probes 199
6.3.2 Arranging Enzymatic Cascades in Vitro 200
6.3.3 Arranging Enzymatic Cascades in Vivo 201
6.4 Challenges and Future Directions 201

7. Nanopatterning for Nanobiotechnologies: Emerging Methods Based on Soft Lithography and Directed Assembly **207**
Aline Cerf, Christophe Thibault, Emmanuelle Trévisiol, and Christophe Vieu
7.1 Introduction on a Vision of the Future of Nanobiotechnologies 208
7.2 Microcontact Printing of Oligonucleotides for Biochip Fabrication 211
7.3 Technological Platform for Molecular Analysis Based on Nanopatterning and Label-Free Optical Detection 216
7.3.1 Deposition of Multiple Probe Biomolecules Using Microcontact Printing and Macrostamps® 218
7.3.2 Diffraction-Based Biosensing Using Nanoscale Periodic Gratings of Probe Molecules 220
7.4 Single Object Investigations Using Soft Lithography 223
7.4.1 First Strategy: Surface Patterning 223
7.4.2 Second Strategy: Use of Structure Topography 225
7.5 Perspectives for Patterning Virus Nanoparticles 230
7.6 Conclusions 231

Index 237

Preface

Enzyme immobilization has over 50 years of well-documented history, with discoveries spanning a large field of applications from sustainable chemistry to medicine through enzyme-targeted therapy. With its increasing understanding of living systems, the scientific community developed a new interest for biologically ordered structures having the potential to become "enzyme nanocarriers" (ENCs).

In nature, proteins interact by means of specific domains. They also combine with lipids and nucleic acids. From these biocomposites originate the three classes of carriers discussed in this book: viruses, polymersomes (inspired from lipidic membrane structures), and DNA origamis. A deep knowledge of the structure of these systems at the atomic resolution and of the physical rules allowing their assembly demonstrated that protein components are key factors for these assemblies. Because of the high power of genetic engineering in designing new protein properties, it has become possible to achieve a high positional control of enzymes on these complex composites. Today, biologists desiring to get into the promising field of bioinspired nanocatalysts have at their disposal a large choice of structures and the tools to modify them.

Many smart applications of these composites have already been proposed. For instance, virus-like particles (VLPs) depleted from viral genetic material can be engineered and used as potent enzymatic nanoreactors or for targeting cytotoxic enzymes to cancer cells.

This book covers some of the most recent advances in this fast-evolving field. We hope that it will help the reader in the conception of new tools to create some ENCs of the future.

Daniela Cardinale
Bordeaux, France
Summer 2015

Preface

[illegible]

Introduction

Enzyme Nanocarriers: What Are They?

Enzyme immobilization on solid supports has been considered for a long time as an attractive solution to perform sophisticated organic synthesis, as required in the preparation of fine pharmaceutical chemicals. A good example is penicillin-G-acylase (PGA), which was immobilized on an epoxidic support. It improved its catalytic features, especially its efficiency in the synthesis of β-lactamic antibiotics and in other kinetically controlled reactions, such as the resolution of racemic compounds. Indeed, PGA is capable of catalyzing acylation reactions far more simply than the chemical methods commonly used in industrial processes. Enzymatic synthesis normally requires mild reaction conditions that are compatible with product stability. The resulting compounds are purer than those obtained by the classic processes of chemical synthesis. Moreover, by avoiding the need to use toxic reagents and solvents, the impact on the environment is greatly reduced. Besides this, the possibility of linking the enzyme with a solid support so as to be able to recover and reuse it effectively renders the study and development of enzymatic processes for the synthesis of penicillin and cephalosporin increasingly interesting for industry. Integration of redox enzymes with an electrode support and formation of an electrical contact between the biocatalysts and the electrode are the fundamental subjects of bioelectronics and optobioelectronics. But with the development of life sciences, chemists became increasingly interested in the performance of enzymes in their genuine environment. In living systems, the cellular organization of cooperating enzymes into supramolecular complexes is a metabolism key feature. A major advantage of such organization is the transfer of biosynthetic intermediates between catalytic sites without diffusion into the bulk phase of the cell. This so-called metabolic channeling combines an enhancement of the catalysis efficiency with a fine-tuning of metabolic pathways. The growing accumulation of knowledge in the field has inspired new technologies to design bioreactors coupling several enzymatic reactions. Consequently considerable effort is produced, aiming at mimicking natural enzyme organization for optimizing the synthesis of valuable metabolites with industrial and medical importance. Today, enzyme immobilization on carriers is considered in several fields, including both fundamental studies covering modern aspects of enzyme kinetics (substrate confinement, limiting diffusion, single-enzyme kinetics, biocatalyzed cascade reactions) and more applied strategies like high-throughput screening of catalytic specificities, bioreactors, sensor ships for analysis, medical diagnosis, and therapy.

In the late nineties, the tremendous progresses of molecular biology fundamentals opened up the possibilities to feed a toolbox for building new bioinspired nanotechnologies. Among them was the goal to reposition biocatalysts in environments mimicking their genuine working place, the cell. The research presented in this book was selected among the most impressive achievements in the field of enzyme bioconjugation to bioinspired nanosupports. It opens up potential applications in nanocatalysis, lab-on-a-chip and biosensor devices, drug delivery vectors, nanometrology, and many more, letting the reader feel that imagination will be the only limitation. We wish this book to be a source of inspiration for the researcher seeking to build smart materials requiring a nanoscale positional control of functional proteins on various carriers. All the supports described herein pertain to soft materials (cells, virus, polymers, DNA). Most of the examples benefit from the amazing properties of proteins and DNA to self-assemble according to the "bottom-up law," a specific feature of all living systems. But these ENCs have the potential to be grafted on solid supports through "top-down technologies," spanning orders of magnitude from the nano- to the mesoscale and above.

Positional Control: A Key

The bottleneck in combining several different enzymes working cooperatively comes from the difficulty to control their relative positional assembly on the support. This control can be achieved by coupling the enzymes of interest with a compatible highly ordered protein scaffold. In the course of evolution, nature has optimized complex architectures conjoining perfect positioning in space with highly specialized functions. These smart materials can be regarded as composites made of a restricted number of building blocks (lipids, proteins, polysaccharides, and nucleic acids). But ultimately, most of these combinations are finely tuned by precise surface overlaps between proteins. The diversity of the combinations of the 20 natural amino acids along the polypeptide sequence results in peptide folding into 3D domains, ensuring a very high selectivity of the assemblies. Virus particles are supramolecular edifices unsurpassed in nature. The simplest of these systems are made of noncovalent combinations of proteins and nucleic acids, which are precisely arranged in space. The information for this self-assembly is programmed in the virus genome through the amino acid sequence of the protein monomers (capsomers) forming the viral particle, or capsid. Although the natural function of capsids is the storage and transport of genetic material, their defined size has made them attractive building blocks from a materials science and nanotechnology point of view. Coupling enzymes to viruses' highly ordered protein backbones is an attractive way to achieve positional control. A very interesting feature of virus scaffolds resides in the diversity of their shapes and physical and chemical properties. Viruses are precious building blocks for nano- and micromaterial design. Their interior space is accessible to small molecules but often impermeable to large ones. Bacteriophages and to some extent plant (cowpea chlorotic mottle virus [CCMV], tobacco mosaic virus [TMV]) or mammalian (hepatitis B virus [HBV]) viruses can serve as good carriers for enzymes. The main advantages are their nanometer size range, their propensity to efficiently self-assemble into monodisperse nanoparticles of discrete shape and size, their stability and robustness, their biocompatibility and bioavailability, and the ease of production in large quantities. In Chapter 1, "Virus Diversity to Explore Various Kinds of Enzyme Nanocarriers," Makinen, Besong-Ndika, and Walter present the vast potentialities offered by virus diversity to design ENCs. Viruses appear to constitute the most abundant and robust biological entities on earth. For instance, the discovery of icosahedral viruses resistant to extreme acidic environments opens up the possibility of using viral architectures for applications that previously seemed unlikely. Ultimately, our creativity might well be the only limitation to the potential use of viral capsids.

Whatever the enzymatic device considered an important issue concerns the topology of enzyme grafting on the support. In an ideal situation, the enzyme must be suitably orientated to facilitate the substrate access to the active site. In Chapter 2, "Strategies for Virus–

Enzyme Coupling," Cardinale, Carette, and Michon review various strategies allowing a positional control of protein grafting inside and outside VLPs using both covalent and noncovalent solutions. The possibility to genetically engineer specific recognition motives on enzymes has opened up new routes to finely tune the orientation of the catalyst on or in the support.

In the cell, the main factor limiting the efficiency of reaction cascades is the Brownian diffusion of reactants. Living systems use essentially two strategies to overcome this limitation. The first strategy consists of the confinement in cellular compartments of the enzymes involved (lysosomes, mitochondria, etc.). The enzymes cannot escape, but these little chambers remain permeable to reactants. Once in the compartment, the substrate is exposed to a high local concentration of enzymes by the confinement effect. The efficiency of catalysis is significantly accelerated compared to reactions in an open medium. In the second strategy, enzymes involved in a given reaction sequence are distributed close enough so as to limit the diffusion. Most of the time, intermediate reactants are trapped in a matrix anchoring the enzymes and diffuse from one catalyst to the other. This matrix can be a phospholipid membrane bilayer, such as the inner membrane of the mitochondria, which contains the adenosine triphosphate (ATP) synthase complex surrounding the respiratory redox cascade. The synthesis of ATP requires a strict positional control of these enzymes related to each other at the nanoscale. In Chapter 3, "Viruses as Model Nanoreactors to Study Enzyme Kinetics," Rurup, Koay, and Cornelissen discuss the use of VLPs' reconstitution to produce a biomimetic confinement of enzymes according to the first strategy described above. A microbial lipase was selectively addressed to the inside surface of a VLP using a peptide-based noncovalent reversible "molecular Velcro" between the coat protein of the virus and the enzyme. In this confined space, the probability of collisions leading to efficient conversion of the substrate is increased. Following this strategy, it is possible to confine cascade reactions catalyzed by coupled enzymes in VLP systems. In addition the authors demonstrate how the limited dilution method allows the confinement of a single enzyme per VLP, providing a tool to study single-enzyme kinetics. Fluctuations in the catalysis rate or velocity distribution of a single enzyme over time (dynamic disorder) cannot be addressed by conventional kinetics. The real-time behavior of a population of isolated enzymes reveals that enzyme conformational dynamics are linked to the catalytic event, an aspect previously only addressed in silico by molecular dynamics simulations. For such studies, the experimental design requires passive adsorption on glass supports or a chemical modification of the enzyme, two processes frequently accompanied by enzyme denaturation. As a solution, VLPs can mimic the small reaction volumes in a cell. The kinetic behavior in real time of a single molecule of horseradish peroxidase (HRP) trapped inside a VLP of CCMV was reported using fluorescence microscopy. The dilution of HRP was adjusted in the presence of a dissociated capsid protein (CP) to generate a mixture of empty VLPs and VLPs containing a single enzyme. When peroxidase was adsorbed to the VLP's outer surface, the product diffused freely and the fluorescence was less localized. In both cases, a significant fluctuation of fluorescence over time was observed.

Alternative Routes to Make Biomimetic ENCs

A system mimicking the natural compartmentalization of enzymes in the cell organelles is described in Chapter 4, "Nanoscale Compartmentalization Techniques in Cascade Catalysis," by van Hest, Willemsen, and Rutjes. This system, defined as a porous polymersome reactor, is based on block copolymers of isocyanopeptides and styrene. Enzymes have been anchored at three different locations, namely, the lumen (glucose oxidase [GOx]), the bilayer (Candida Antarctica lipase B [CalB]), and the surface (HRP) of the polymersomes. Such an approach, applying the most sophisticated modern chemistry, allowed the authors to demonstrate that it is possible to control an enzymatic cascade of a reaction using a differential compartmentalization of the biocatalysts. This book is not intended to be exhaustive in the field of ENCs. Hence, we focused on a selection of approaches illustrating the most significant advances in the field. The reader will have noticed that however elegant and sophisticated they may be, these achievements mainly aim to be a proof of concept. However, they already participate in a box of molecular tools that technologists have begun to use. In terms of promising applications, we could not omit the use of ENCs in the medical field. Chapter 5, "Nanocarriers for Therapeutic Enzymes," treated by Howard, Hood, and Muzykantov, focuses on different concepts supporting these applications.

Assessing the effect of spatial organization on enzymatic activity in multienzyme systems is of primary interest. It will not only help to understand the molecular mechanisms involved but also will have a high impact for the design of nanotransducers and lab-on-a-chip applications. To this aim, nanoscale experimental platforms need to be designed. There are very few methods available to systematically evaluate how spatial factors (e.g., position, orientation) influence enzymatic activity in multienzyme systems. This limitation notably comes from the fact that their small size makes it extremely difficult to organize biomolecules onto surfaces in order to form fully active supramolecular complexes amenable to experimental studies. Recent advances in DNA–protein conjugates make the absolute control of enzymes positioning on well-defined 2D DNA origami possible. This provides a true genetically programmable bottom-up self-assembly of enzyme complexes on DNA templates. Using a chemical conjugation of enzymes and DNA staples specifically hybridizing with topologically defined sequences on the 2D DNA supports, the catalytic efficiency of the cascade can be finely tuned as a function of the distance between enzymes, so demonstrating the channeling effect discussed above. However, enzyme orientation on the support remains uncontrolled because of conjugation sites all around the protein surface. Far more sophisticated is the use of a whole biological organism as microplants to build networks of enzyme cascades on RNA templates. Such an approach belongs to the very active field of synthetic biology. Bacteria were genetically programmed to produce RNA self-assembling in large areas displaying anchoring sequences for small RNA aptamers allowing specific binding to the enzymes composing the cascade. It was demonstrated that the efficiency of the cascade is modulated by the assembly geometry [1]. In Chapter 6, "DNA Origamis as Protein Nanocarriers," Elezgaray, Aimé, and Arbona, after a general introduction on DNA origamis, present the potential of 2D and 3D DNA constructs as carriers. Although representing a remarkable experimental achievement, this approach has, up to now, presented notable drawbacks for "out of the cell" applications. This supramolecular organization of enzymes using a totally bottom-up approach is difficult to extend from nano- to mesoscopic-length scales. Moreover, in terms of applications, DNA-based structures are sensitive to temperature and many other physical and chemical parameters. However, very recent works demonstrate that such difficulties are progressively overcome. In particular, the organization of origami units at the mesoscale and above could be accessible in the near future [2]. Chapters 3–6 deal with nano-object carriers, which can be used for the transport of enzymes. It was discussed above that, because of their highly ordered nature, virus-like structures can be precisely decorated with enzymes and can be used as ENCs. It appears that ENCs are easier to position on a support than single enzymes using top-down processes such as nanolithography or convective-capillary deposition. In Chapter 7, "Nanopatterning for Nanobiotechnologies: Emerging Methods Based on Soft Lithography and Directed Assembly," Cerf, Thibault, Trévisiol, and Vieu present the last developments of top-down technologies enabling a precise patterning of single nano-objects such as virus particles or DNA molecules on various supports. This illustrates how bottom-up and top-down approaches begin to meet for the preparation of smart materials. This approach aims at bridging the gaps between the mesoscale, the microscale, and higher.

The interest for virus-based technological applications increases in a variety of emerging fields from self-assembled nanoscale computers and machines to drug delivery vehicles, biochips, and self-healing regenerative tissues. In addition, alternative enzyme nanocarriers such as polymersomes and DNA are under investigation. This field of expertise will become increasingly important in the coming decade.

This edition of the book gets now to completion but science is still running. Recent developments in synthetic biology have led the concept of ENCs much farther. Bacterial cells can now be redesigned for the in situ assembly of enzymes networks on RNAs origamis. An actualized table content of the book should certainly include a very last chapter from a team such as the one of Delebecque et al. [1] whose work allows to envisage the engineered bacterium itself as a true sensor chip.

Thierry Michon
Bordeaux, France
Summer 2015

---

1. Delebecque, C.J., Lindner, A.B., Silver, P.A., and Aldaye, F.A. (2011). Organization of intracellular reactions with rationally designed RNA assemblies. *Science*, 6041, 470–474.

2. Teshome, B., Facsko, S., and Keller, A. (2014).Topography-controlled alignment of DNA origami nanotubes on nanopatterned surfaces. *Nanoscale*, 6(3), 1790–1796.

Acknowledgment

The editors are grateful to the EU, which, in the frame of the cost action "BioInspired Nanotechnologies: From Concepts to Applications," 2010–2013, supported a fruitful networking between the contributors of this book. They wish to thank the French Agence Nationale pour la Recherche (Viruscaf CP2N 2009 grant and Cascade Piribio 2009 grant).

Chapter 1

Virus Diversity to Explore Various Kinds of Enzyme Nanocarriers

Jane Besong-Ndika,[a,b] Jocelyne Walter,[b,c] and Kristiina Mäkinen[a]
[a]*Department of Food and Environmental Sciences, P.O.Box 27, 00014 University of Helsinki, Finland*
[b]*Equipe de Virologie, UMR 1332 Biologie du Fruit et Pathologie, INRA et Université de Bordeaux, CS 20032, 33882 Villenave d'Ornon Cedex, France*
[c]*CNRS, Délégation Aquitaine, esplanade des Arts et Métiers, BP 105, F-33402 Talence Cedex, France*
kristiina.makinen@helsinki.fi

1.1 Introduction to Virus Diversity

Viruses are organisms that use the host cell as a factory to multiply and assemble vast amounts of progeny virus particles. The virus particles enclose viral genomes to provide them with protection and a mode of transport to new host cells. The protein shell surrounding the nucleic acid genome consists of capsid (or, as a synonym, coat) protein (CP) subunits. The virus capsid is often seen as a "protein shell or tube surrounding the virus genome," but in fact in the cellular context it represents a complex and dynamic macromolecule

Enzyme Nanocarriers
Edited by Daniela Cardinale and Thierry Michon
Copyright © 2016 Pan Stanford Publishing Pte. Ltd.
ISBN 978-981-4613-42-2 (Hardcover), 978-981-4613-43-9 (eBook)
www.panstanford.com

responding to various cellular and environmental signals [1]. CPs are multifunctional proteins, and they can be involved at almost every stage of the viral infection cycle [2].

The simplest virus particles consist of hundreds to thousands of identical subunits of CPs arranged in one or a limited number of symmetric patterns to create a minimum-energy structure (Fig. 1.1). Depending on the virus species the structural units from which capsids are formed may as well comprise different protein subunits. The first structural information on viruses was on tobacco mosaic virus (TMV; genus *Tobamovirus*; reviewed in [3]), revealing architecture based on helical symmetry. Other commonly used symmetries underlining the structures of virus particles are icosahedral symmetry or a symmetry consisting of a combination of a helical and an icosahedral unit. Diameters of icosahedral virus particles range from 17 nm to 800 nm and the sizes of filamentous viruses from 300 nm to over 2000 nm in length and from less than 10 nm up to 80 nm in width. The size of the virus particle is in general proportional to the size of the genome it encapsidates [4]. Complex virus structures may contain lipid envelopes consisting of a lipid bilayer carrying viral proteins, portal complexes, and core capsid layers. A preformed particle structure may be modified by

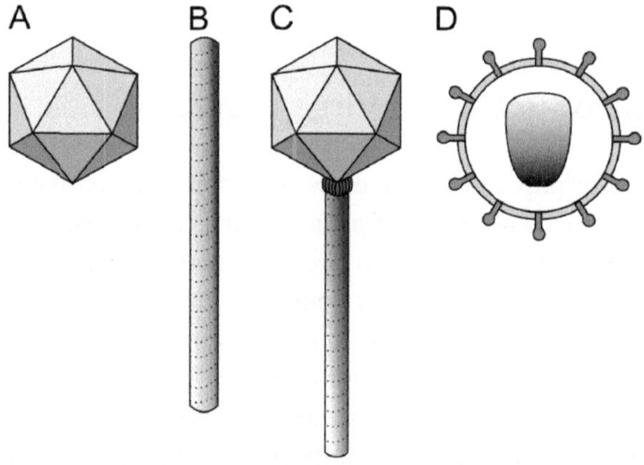

Figure 1.1 Schematic presentation of virion morphotypes. (A) Icosahedral symmetry. (B) Helical symmetry. (C) Combination of icosahedral and helical symmetries. (D) Enveloped asymmetric virion. The virions are not in scale. Based on Ref. [6]. Reproduced from the PhD thesis of Maija Pietilä (ISBN 978-952-10-8563-5; ISBN e-thesis 978-952-10-8564-2) with permission.

maturation upon proteolysis and conformational changes. Some enveloped viruses have pleomorphic structures (Fig. 1.1), and especially archaea-infecting viruses may be bottle, lemon, or spindle shaped [5]. The regular nanoscale virus structures made by nature have inspired researchers to utilize them in nanobiotechnology applications.

Viruses are numerous in all environments. The estimated number of virus particles in the biosphere at any given time is 10^{31} [7]. Among these roughly 95% are bacteriophages. The International Committee of Taxonomy of Viruses (ICTV) has classified the virosphere on the basis of the host type and the nature of the genome currently into 2284 virus and viroid species in 349 genera, 19 subfamilies, 87 families, and 6 orders (Fig. 1.2; [6]). Taken together the amount of genetic information stored in the genomes of viruses is vast and yet to be explored. Remarkably less variety can be found in the architectures of virus particles. The first structural studies leading to high-resolution (2.9 Å) data on tomato bushy stunt virus (TBSV; genus *Tombusvirus*; [8]) and southern bean mosaic virus (SBMV; genus *Sobemovirus*; [9]) capsids revealed an eight-stranded, β-barrel CP fold. A similar fold was later found from, for example, the common cold virus [10] and poliovirus (PV: genus *Picronavirus*; [11]). The accumulation of structural data on virion architectures has made it possible to perceive that many virus groups and families can be organized into few major structure-based lineages according to a similar major capsid protein (MCP) fold [12–16]. Thirty-three of the assigned eighty-seven families in the ICTV virosphere can be tentatively placed into one of these four lineages, while twenty-two of the virus families have been assigned to an order (Fig. 1.2B; [16]). The limited number of architectures utilized by viruses may reflect the fact that only a subset of protein folds suffices as building blocks of virions.

Viruses can be produced in high quantities. They are easily purified, biocompatible and biodegradable substances with a high degree of symmetry and polyvalency and a low degree of polydispersity. On the basis of these features, viral nanoparticles (VNPs) have a great potential to be exploited as functional devices that can be applied in fields such as biosensors and coatings for biotechnical and medical applications. The wide range of inherent properties of virus particles provides an enormous variety of potential application platforms. VNPs, which contain the viral genome, are infectious, but the properties of some CPs allow self-assembly in the absence of viral RNA. These particles are called virus-like particles (VLPs). VLPs/modified VLPs can in many cases tolerate extreme temperatures [17, 18] as well as various solvents [19]. When compared to infectious VNPs, VLPs can be considered preferable from the biosafety point of view.

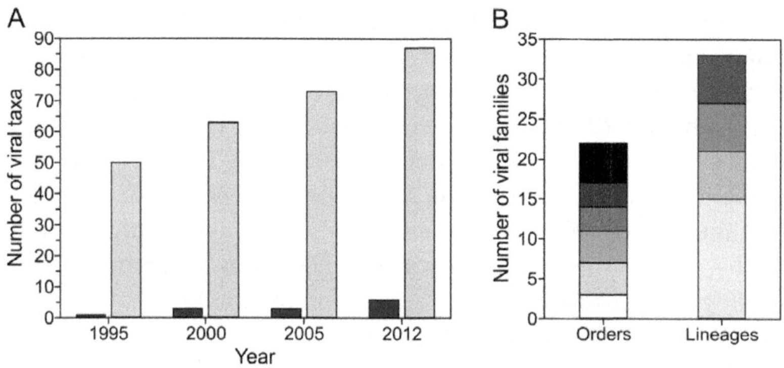

Figure 1.2 Virus classification. (A) Increasing number of viral orders (dark gray) and families (light gray). (B) Viral families assigned to six orders, *Caudovirales* (3 families), *Herpesvirales* (3), *Mononegavirales* (4), *Nidovirales* (3), *Picornavirales* (5), and *Tymovirales* (4), and the four lineages, BTV-like (6), PRD/adeno (6), HK97-like (6), and picorna-like (15), are indicated in different shades of gray. The given number indicates the number of viral families assigned to each order or lineage. Based on Refs. [6, 16]. Reproduced with minor modifications from PhD thesis of Maija Pietilä (ISBN 978-952-10-8563-5; ISBN e-thesis 978-952-10-8564-2) with permission.

The repetitive nature of the protein subunits in virions provides a chemically uniform platform for selective attachment and presentation of chemical and biological moieties. Many chemical strategies have been developed to modify VNPs and VLPs for ligand attachment, immobilization to surfaces, and formation of complex aggregates and networks [20]. CPs are also amenable to genetic modifications as long as the peptide or protein fusion does not disturb their ability to self-assemble. Information on structural, chemical, and physical properties of virus particles is a prerequisite for their exploitation as VNPs. In this chapter entitled "Virus Diversity to Explore Various Kinds of Enzyme Nanocarriers," the different biological properties contributing to the formation of various types of VNPs and specifically viral enzyme nanocarriers (ENCs) are discussed.

1.2 Assembly and Packaging Strategies of Viruses

Virion components, including viral genome, CPs forming the coat around the genome, and membranes associated with nucleocapsids (NCs), come together in a process called assembly. The self-assembly process of viruses was first demonstrated by Fraenkel-Conrat and Williams in 1955 [21]. This study showed that disassembled TMV particles could be reconstituted in vitro and the particles maintained their infectivity. Virion assembly occurs basically with two mechanisms: (i) They are coassembled with the genome, or (ii) viral genomes are packaged into an empty preformed procapsid in a process powered by nucleoside triphosphate (NTP) hydrolysis.

The assembly process is driven by the laws of thermodynamics, meaning that the subunits forming the particle are energetically in a more favorable environment within the virion than the free subunits [22]. The structure of icosahedral virus particles can be described on the basis of the triangulation number defining the possible icosahedral surface lattice designs, quasi-equivalence describing the nearly equivalent bonding, and self-assembly. As many copies of one protein compose the particle the required information for self-assembly resides in the specific bonding patterns of the individual proteins, although with some viruses the assembly process is regulated by scaffolding proteins that are not present in the final virion [23]. Cementing or glue proteins may exist in final virions to give stability, and some other proteins may be incorporated into virions to serve specific functions like binding of the virus to its receptor on the cell surface.

To fully understand the assembly mechanism it is important to know the composition of the virion as well as all intermediates formed during the assembly pathway. The biology of virion assembly as well as the requirements to form stable mature particles set boundaries for the doable genetic and chemical modifications of the virions. Therefore it is especially important for ENC applications to solve the surface exposed areas that do not take part in CP interactions needed in the assembly pathway or for the stability of the mature virions.

1.2.1 Particle Formation via Coassembly

The capsids of helical viruses, both rod-shaped and flexuous filaments, and small icosahedral RNA viruses are coassembled with the genome. Helical viruses are viruses that have their RNA enclosed in a protective shell consisting of identical protein subunits arranged in a helical manner. This basic structure was first proposed by Crick and Watson in 1956 [4]. Their studies led them to conclude that the structure of small viruses consisted of identical units with specific bonding properties arranged to form a close surface around the viral genetic material, the main events being nucleation and elongation in the order mentioned.

Virus assembly and stability of virus particles is governed by several molecular interactions. Viruses of the genera *Tymovirus* and *Comovirus*, for example, are stabilized mostly by protein–protein interactions and hence can form VLPs in the absence of viral RNA, while viruses of the genera *Alfamovirus, Bromovirus*, and *Cucumovirus* are primarily stabilized by RNA–protein interactions and require other virus-encoded proteins for successful packaging [24, 25]. In spite of these two being the major acting forces, other sequence-dependent/independent RNA–protein interactions as well as structure-dependent interactions are also important. One amongst other constraints in coassembly of icosahedral viruses is the dimension of the capsid, which inflicts a restriction on the size of the RNA to be encapsidated. Studies on turnip crinkle virus (TCV) by Qu and Morris exemplify this aspect [26]. The filamentous virus structure is more flexible in this respect, as the virion size can increase with increasing genome size, as shown with potyviruses [27]. Next we attempt to picturize the assembly process of small simple viruses using TMV as a model to enlighten the major steps.

1.2.1.1 Assembly of TMV

TMV is a rod-shaped RNA virus, 300 nm in length, 18 nm in diameter, and a central radius of 2 nm [28]. Its virus particle contains about

2100 protein subunits enclosing a single RNA molecule. Hence, RNA makes only 5% of the particle, while 95% contains the protein shell. The length of the protein shell seems to be dependent on the length of the viral RNA. Intensive and informative studies on the in vitro assembly process enabling the proper understanding and uncovering of the TMV assembly have been communicated. These studies are used here as a prime example of the simple assembly process involving condensation of genomic single-stranded RNA (ssRNA)(+) and CPs. The major component of the TMV rod is the 17 kDa CP comprising 154 amino acids. Depending on certain conditions, pH being the most important of all, protein monomers constituting the virion can self-assemble to form different aggregates (Fig. 1.3).

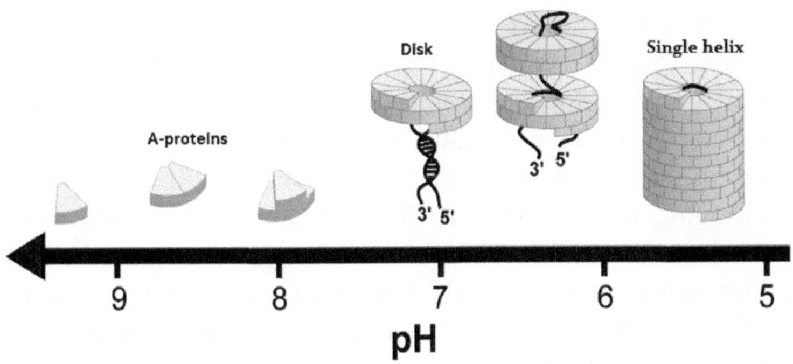

Figure 1.3 Nucleation and elongation during the TMV assembly process. Elongation is immediately followed by addition of single-protein subunits at the 3′ end of the RNA and coat protein disk at the 5′ end.

1.2.1.1.1 *Nucleation event*

In the presence of TMV RNA, the first determining step in the assembly process is the nucleation event, which is characterized by the initial interaction that occurs between the CP double disk and the specific viral RNA signal [29, 30]. This necessitates 17 protein subunits that align themselves along an RNA molecule, forming a complete turn. This process is kinetically very slow. Butler and Klug in 1971 showed that the rate of the assembly could be increased significantly by starting the process with disk preparations [31]. The TMV CP disk interacts with a specific sequence in the RNA molecule. This sequence can be termed “origin of assembly” (OS), or packaging signal (PS) or encapsidation signal (ES). In TMV this sequence is about 65 nucleotides long, located between nucleotides 5420 and 5546 at the 3′ end of the TMV RNA [32–34]. The core sequence, 5′-AAGAAGUC(U)G-3′, forms a stem-loop (SL) structure and is sufficient to initiate the assembly if placed into a heterologous RNA molecule [35, 36]. Each protein subunit is said to bind three nucleotides. Nucleation leaves the 5′ end of the RNA looped back down the central hole of the disk, while the 3′ end extends freely from the bottom of the helix [37, 38].

ESs of many other viruses have been identified. In Table 1.1 are documented examples of the ES of a few other ssRNA(+) viruses, which also reveal the importance of the secondary structural element of their RNAs on nucleation. In hepatitis B, ribosomes translating this ES sequence from the precore start codon are essential for packaging [39]. Some viruses have *cis*- and *trans*-acting factors that assist in packaging. Brome mosaic virus (BMV), for example, contains a *cis*-acting packaging element (PE), also essential for packaging, in addition to its 3′-tRNA-like structure [40]. Other viruses, for example, Ross River virus (RRV), possess more than one ES along their viral genome [41]. It is fair to surmise that despite the importance of the ES, its presence alone might not be sufficient for packaging due to other acting factors like secondary structures, *cis*- and *trans*-acting factors, and other elements on specific positions on the genome. The nucleation event is also very dependent on the protein subunit with specific RNA-binding activity. An arginine-rich RNA-binding motif (ARM) of the CP, which interacts with viral RNA, is important for genome packaging of viruses of the genera *Bromovirus* and *Sobemovirus* [42–44]. This ARM, which exhibits different conformations, has been made evident in other viruses such as human immunodeficiency virus (HIV) and bacteriophage λ and has been shown to be important for interactions with viral RNA, specifically or nonspecifically. Deletion of the first amino acids of BMV impaired assembly of the virus in vivo [45]. In 2004, Linger et al. [46] determined a 32-amino-acid peptide capable of recognizing the Sindbis virus (SINV) ES RNA, which upon binding induces a conformational change on the RNA. It could be assumed that interaction of these protein subunits with the RNA during the nucleation event can as well induce specific folding activity of the

RNA. Karran et al., in 2011, revealed a new level of favoritism by the CP whereby it was able to recognize specific changes in RNA at the level of bases and reject this RNA for packaging [47].

Table 1.1 Critical loops on the RNA secondary structures of some (+) ssRNA viruses that act as signals for viral RNA encapsidation by protein subunits

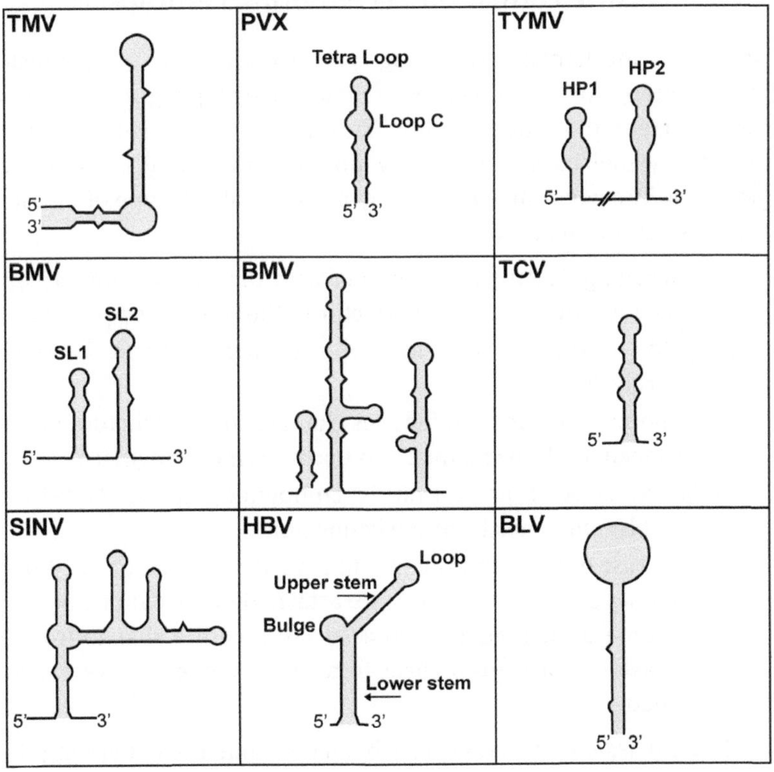

1.2.1.1.2 *Elongation*

Elongation of the rod by addition of CP aggregates to completely coat TMV RNA seems to proceed in a bidirectional manner. Lomonossoff et al. in 1979 [48] and Fairall et al. in 1986 [49] both demonstrated this was a simultaneous process with both the 5′ end and 3′ end being encapsidated at the same time, even though coating occurred faster toward the 5′ end. Faster growth of the rod was observed when the proteins were added as disk structures. Several studies on the mechanism of elongation of the 3′- and 5′-RNA ends have brought out discrepancies in both methods. Elongation at the 3′ end requires single-protein subunits of A-protein, being an assembly of three or more protein subunits [48, 50], while elongation at the 5′ end requires disk-like structured aggregates [51].

1.2.2 Particle Formation Via Preformed Procapsids

Assembly mechanism requiring preformation of a procapsid is exploited, for example, by tailed bacteriophages [52] and herpesviruses [53] having double-stranded DNA (dsDNA) genomes and by cystoviruses [54] having double-stranded RNA (dsRNA) genomes. The generalized assembly pathway of this type of viruses involves the following steps:

(1) Packaging ATPase in a complex with the DNA genome interacts with the head–tail connector of the preformed prohead.
(2) DNA is translocated into the prohead driven by ATP hydrolysis.
(3) Genome translocation induces changes in the structure of the prohead. It often expands and becomes more angular.
(4) When DNA translocation is completed, the ATPase–DNA complex detaches from the connector.
(5) The rest of the components, like neck, tail, and/or tail fiber, yielding an infectious virus are attached to the connector site. Genome packaging is often associated with a change from a metastable procapsid form to a more stable mature capsid structure.

In vitro assembly studies have been instrumental in defining the components and the order in which they are incorporated. Such a study revealed intermediates and conformational changes along the assembly pathway of the cystovirus phi6 [55]. It was established that formation of an empty procapsid requires the shell-forming protein P1, viral RNA-dependent RNA polymerase P2, packaging NTPase P4, and a procapsid-stabilizing protein P7 (Fig. 1.4).

These in vitro assembled dodecahedral polymerase complexes (PCs) were able to bind and package the genomic s, m, and l ssRNA precursors and to replicate them into dsRNAs. The mature dsRNA-filled PC becomes spherical. Packaging of the viral RNAs relies on encapsidation and replication signals found from the 5′ and 3′ ends of genomic RNA molecules. Assembly of the NC surface protein P8 into full procapsids led to the formation of NCs. The next intermediate was found to be formed by proteins P6, P9, P10, and P13 and lipids forming an envelope. Finally, virions attained infectivity upon acquisition of spikes P3. A detailed understanding of the assembly mechanism of phi6 PCs and the knowledge about the signal RNA sequences required for encapsidation have opened up fascinating possibilities to utilize the enzymatic activities associated with phi6 PCs for heterologous dsRNA synthesis, as will be explained later in the chapter.

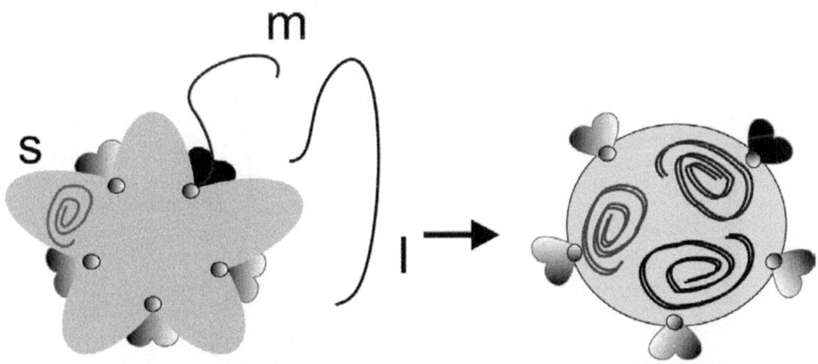

Figure 1.4 Phi6 PC is an intermediate of the Phi6 assembly pathway. Phi6 PCs have been utilized as nanodevices to produce heterologous dsRNA molecules.

1.3 Virion Formation and Its Multiple Roles in Virus Infection

The architecture of the virion must provide an optimal interaction dialogue with the host machinery to start a successful infection (Fig. 1.5). The architecture of the infectious virion must therefore have enough flexibility to allow disassembly and release of the infectious genome. For any given final architecture of viruses, assembly of infectious virions is a specific and highly regulated process requiring specific protein–protein and protein–nucleic acid interactions.

Interactions between CPs and viral RNA may lead to either inhibition or enhancement of viral gene expression. Changes in CP

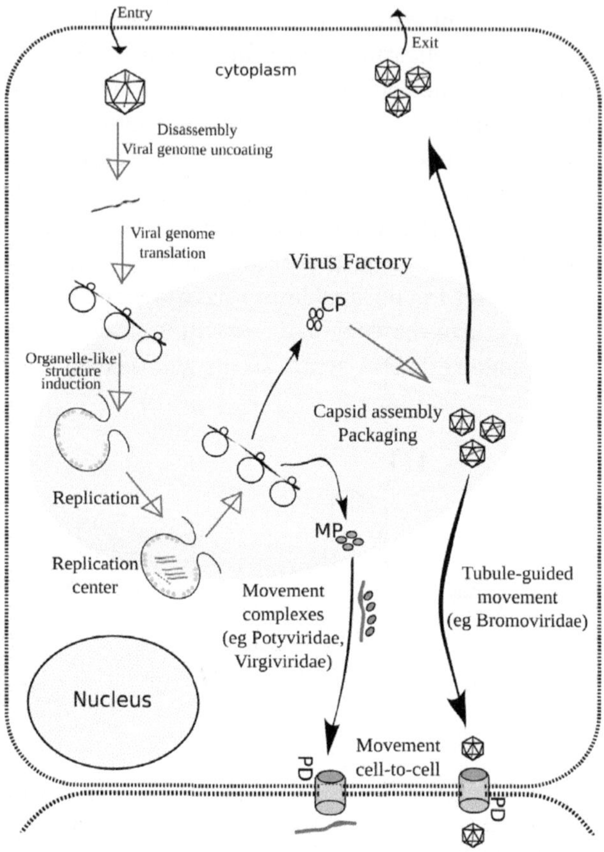

Figure 1.5 Virus replication cycle example of plant viruses that replicate in the cytoplasm. Upon entry into the cell (such as through insect vectors) disassembly of the capsid allows viral genome uncoating and translation of viral replicase proteins. Genome replication is compartimentalized in organelle-like structures (e.g., vesicles, spherules, multivesicular bodies) that enhance replication efficiency and protection from host defenses. Neosynthesized viral RNAs egress into cytoplasm where translation of viral factors such as movement protein (MP) and coat protein takes place. Virus factories may assemble in a large and sophisticated network of connecting membranous structures that represent adapted biological structures not only for genome synthesis but also for the other steps of the virus life cycle, such as virion assembly, cell-to-cell movement, and exit of the cell. Viral progeny is transported into the new cell, either in the form of a ribonucleoprotein complex or as an intact virion (for reviews see [56–61]).

levels may regulate the progression of the viral infection cycle from genome replication/translation to virion assembly [62]. In several plant virus genera, assembly into an infectious virion is highly crucial because it is necessary for cell-to-cell and long-distance movement as well as for acquisition and dissemination to new hosts by a wide range of vectors (e.g., aphids, nematodes, fungi) [56, 63].

Investigations on the mechanisms involved in disassembly of TMV, which is a rod-shaped positive-stranded RNA virus, allow us to propose the following scenario: When entering the cell, TMV particles undergo cotranslational disassembly during which CPs are removed from the part of the particle bearing the 5′ end of the RNA as the first reading frame is being translated [1, 64–66]. The molecular mechanisms involve disruption of carboxyl–carboxylate interactions between neighboring subunits and of phosphate–carboxylate interactions between CPs and RNA due to reduction in calcium and proton concentrations that viruses encounter in a cytoplasmic environment [66].

Icosahedral plant viruses with positive, ssRNA of different genera—BMV (genus Bromovirus); cowpea chlorotic mottle virus (CCMV; genus *Bromovirus*), and SBMV—undergo structural transition induced by pH and calcium swelling, after which the genomic RNA becomes accessible to ribosomes. In the case of BMV, the proposed scenario would involve a major structural transition of the *N*-termini of the five subunits of the pentameric capsomere from the interior to the exterior of the virion. The RNA would become accessible for translation by means of this induced channel [66].

The dsDNA (RT) genome of cauliflower mosaic virus (CaMV; genus *Caulimovirus*) and the ssDNA of Geminiviridae have to be transported to the nucleus, where they are transcribed to mRNA by the host RNA-dependent RNA polymerase (RdRp). Their CPs have a nuclear localization signal (NLS), which may target the particle into the nucleus [66, 67]. As for plant viruses, animal viruses that replicate in the nucleus need to deliver their genome into the nucleus, therefore housing the NLS in their CP [68]. The crossing of a nuclear membrane uses a nuclear pore complex (NPC), occurs in several ways, and is associated with various levels of capsid disassembly, since viruses may bypass intact (e.g., parvoviruses and polyomaviruses) or, as is the case for herpesviruses, only the viral DNA genome is ejected into the nucleus [69–71]. Studies on enveloped and nonenveloped animal viruses indicate that uncoating occurs through complex, multistep processes triggered by virus–host cell interactions, highlighting the involvement of different factors during uncoating (for reviews, see [72, 73]).

Capsid assembly and viral genome packaging are highly dynamic processes coordinated with other steps of the virus life cycle, and they can take place either in nuclear or cytoplasmic compartments of the cell [56, 66, 70, 71]. In eukaryotic hosts, many viruses assemble their capsid around their genome in the cytoplasm. Some viruses replicating in a nucleus assemble their capsid in virus factories into the nucleus (e.g., Adenoviridae, Polyomaviridae, and Herpesviridae); thus all capsid components are targeted to the nucleus through the NLS-containing proteins. Viral capsids assembled in the nucleus are released into the cytosol by hijacking an endogenous cellular pathway that serves to transport cellular cargo exceeding the size limit imposed by the NPC [70, 71].

In the cell, viral genome packaging must differentiate between the host and the viral nucleic acid, suggesting a fine-tuned packaging of viral progeny genomes. The main strategy for packaging initiation relies on the recognition of an interaction between the structural CP and a specific RNA signal [74, 75–77]. Genome packaging is functionally coupled to replication in RNA viruses pathogenic to humans (PV), insects (flock house virus [FHV]), and plants (BMV). The requirements for packaging BMV RNAs are more complex than electrostatic interactions requirements [40, 78]. During infection, only replicated BMV RNAs are suitable for encapsidation and efficient packaging of RNA4 is coupled, not to its transcription, but to the translation of the CP from RNA4. The BMV virion can also form a pseudo $T = 2$ particle when BMV RNA is absent, suggesting that the particles have a mechanism to form preassembled immature particles until RNA can be encapsidated [79]. BMV RNA3 can also be encapsidated without RNA4, suggesting the existence of a mechanism to bring RNA4 into a particle that is at least partially assembled [78, 80]. Among the BMV *cis*-acting RNA sequences, a putative packaging specificity signal in RNA3 has been identified [62]. An update on the key role of BMV CPs in orchestrating the regulation of RNA synthesis, translation, and RNA encapsidation is reviewed in Kao et al. [62]. Investigations on FHV highlight that a physical interaction between viral replicase and CPs is required for genome-packaging specificity [81]. The involvement of replicase in packaging has been demonstrated in animal viruses such as PV [82], FHV [83], and Kunjin virus (KUNV) [84]. Investigations suggest that for plant viruses, too, in addition to its role in the synthesis of progeny RNAs, replicase may function as a "specificity filter" by blocking host RNA packaging and maintaining physical homogeneity among progeny virions [74].

Among helical ssRNA(+) genome viruses, rattlesnake-like virions in the Closteroviridae family (genus *Ampelovirus, Closterovirus* [monopartite genome], and *Crinivirus* [bipartite genome]) are long, flexuous virions with a unique bipolar architecture. The virion body (95% of the virion) is assembled by the MCP and the tail found at one end of the virion by the minor capsid protein (CPm). Two additional virus-encoded proteins, an HSP70 homologue and a 61 kDa protein, are tightly attached to this tail structure, which is required for virus movement [85]. Similarly, plant potyviruses contain such a structure at one end of the virion [86]. Also, such a specific structure of the virion may be an attractive option to be engineered to carry enzymes.

As shown above, the formation of virions in the infection context is an interplay of many factors and functions, and for this reason the use of infectious virions in ENC technology may be unnecessarily complicated. Therefore, for most VNP and ENC applications production of VLPs is more suitable.

1.4 Production of VLPs

VLPs, as mentioned earlier, allude to protein complexes formed from the self-assembly of viral structural proteins in an attempt to mimic the native viruses. Self-assembly refers to the ability of the constitutive protein subunits of a virus particle to associate under appropriate conditions into a macromolecular structure in the absence of RNA. Structure-wise, they can be similar or not similar to the native virus. The assembly of this protein subunit is greatly reliant on pH and ionic strength (IS).

1.4.1 Helical Viruses

In TMV, for example, pH plays an important role in the stability of its VLPs. At neutral pH and a long range of IS, it aggregates [87] into a two-layer disk, which forms the basis of the formation of the rod-shaped particle. These disks can also be formed from an association of A-proteins in a pH 8–9 solution. This disk also plays an important role during the nucleation event at the beginning of the assembly process [88, 89]. At pH below 7, this disk associates in a polar manner, forming a helix as a consequence of hydrophobic interactions [29, 90]. The negative charges on the carboxylic side chains of Glu 50 and Asp77 in the CP at low pH repulse each other, leading to disassembly of this virus [91]. These TMV VLPs are very similar to the TMV virus, with the only difference that they might contain 161/3 or 171/3 subunits per turn [92], while the virus contains only 161/3 subunits.

1.4.2 Spherical Viruses

For small icosahedral viruses pH is not only crucial for assembly and disassembly but also induces a conformation change on the virus structure, as shown in Fig. 1.6 [93, 94].

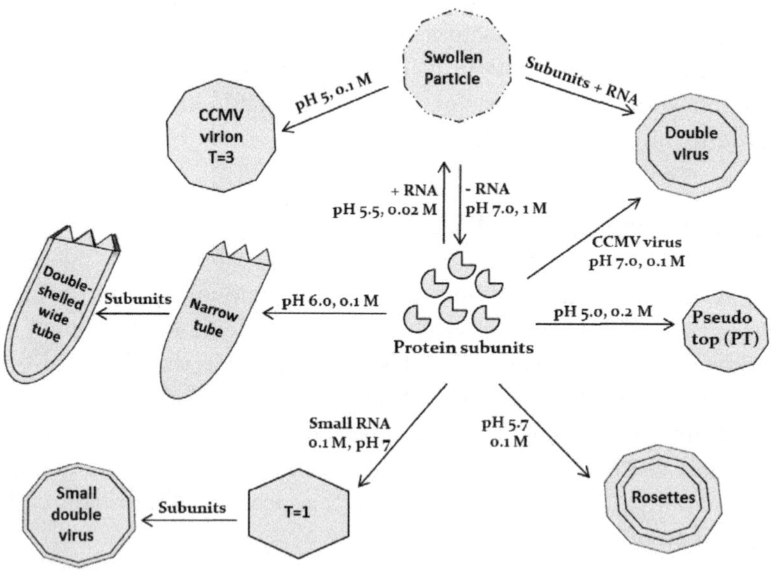

Figure 1.6 Effect of pH and ionic strength (NaCl) on the assembly/disassembly of CCMV CP subunits and formation of chelate-like structures in the presence or absence of RNA.

CCMV (genus *Bromovirus*), the first icosahedra virus of which the assembly was studied, swells to form tube-like structures instead of the usual $T = 3$ structure at physiological pH. This is a consequence of the weakening of protein–protein interactions, and at this stage only protein–RNA interactions hold the structure together [94]. This effect is also observed with BMV and broad bean mottle virus (BBMV) [93]. This swelling can be reversed at pH 5; however, infectivity is only regained in the presence of Mg^{2+} [95, 96]. Swelling could be the result of carboxyl groups, which ionize at pH 7.0 and repel each other [94]. Mg^{2+} could stabilize the virus; BMV is stabilized by Mg^{2+} and Ca^{2+} [97]. Treatment of these swollen particles with nucleases resulted in the formation of chelate-like structures such as long rounded tubes, small icosahedra particles, ellipsoidal particles, and double-shelled particles probably due to the dissolution of the RNA secondary structure. Table 1.2 enumerates the conditions for assembly and disassembly of some VLPs, with or without the presence of RNA, and structure-wise comparisons to the native virus are made.

Table 1.2 Conditions for assembly of some VLPs, with or without the presence of RNA, and structure-wise comparisons to the native virus are made

	Assembly	Disassembly	VLP versus virions (structure-wise)	References
TMV Helical	Acidic pH (5–6) High *T*° HIS	Alkaline pH Low *T*° LIS	Native: 161/2 subunits per dick turn. VLP: 161/2 or 171/2 subunits per disk turn	[29, 92]
TCV Sphere	pH 7–7.8 EDTA/$CaCl_2$ LIS	Alkaline pH 0.5 M IS Low *T*°	Native: $T = 3$ particles VLP: $T = 3$, $T = 1$, ellipsoidal particles, larger $T = 3$, $T = 1$ particles	[98, 99]

(*Continued*)

Table 1.2 (*Continued*)

	Assembly	Disassembly	VLP versus virions (structure-wise)	References
PVX Helical	1:20 RNA: CP ratio 2 mM NaP pH 6.2 RT	0.1 M IS Low *T*° pH 7–8	Native, VLP: Structurally similar	[100, 101]
PVY Helical	1–100 mM PB pH 6–9	pH 5–8, 10.5 2 M IS	Native: Pitch of 3.3 nm VLP: Pitch of 4.0 nm, particles containing stacked disk	[101, 102]
CCMV Sphere	10 mM Tris 10 mM KCL 5×10^{-3} M $MgCl_2$ 10^{-3} ClR 4°C	1 M NaCl 20 mM Tris 10^{-3} M ClR pH 7.4 4°C	Native: $T = 3$ VLP: Smaller $T = 3$ At pH 7.0, chelate-like structures formed	[103, 104]
RNA Phages MS2	20 mM Tris-acetate, pH 7.0–7.5 1 hr on ice Short RNA	Glacial acetic acid	Native, VLP: Similar $T = 3$	[105]
RNA phage PP7	50 mM Tris-HCl, pH 8.5 RNA	No information found	Native: $T = 3$ VLP: More stable $T = 3$ stabilized by interdimer disulfide bonds	[106]
HIV-1	0.02 M Tris, pH 8.0 0.1 M NaCl 5 mM β-ME RNA	No information found	No suitable comparison found	[107]
XMRV	50 mM Tris buffer, pH 8.0, 100 mM NaCl 1 μM $ZnCl_2$ DNA/RNA 4°C	No information found	Native, virions: Similar	[108]

1.4.3 VLP Expression Systems

VLPs are all produced either by expression of the structural protein(s) with direct purification of the assembled particle or via in vitro assembly after purification of the required structural protein(s). Several host systems have been used for the production of VLPs in vivo using expression vectors particularly adapted for large-scale protein production. Some include bacterial, insect, yeast, plant, and mammalian systems. All these systems have their advantages and disadvantages in VLP production and depending on the application can be used appropriately.

- *Bacterial systems*. Sugarcane streak mosaic virus, cardamom mosaic virus, Johnsongrass mosaic virus, and many other examples of potyviral CPs are documented to form VLPs in vivo in bacterial expression systems [109–111]. The setbacks with using bacterial systems are their inability to undergo post-translational modifications (PTMs), their inability to produce soluble proteins in most cases, degradation by bacteria proteases, codon bias, presence of toxins, which could be the consequence of low yield, and heterogeneous/contaminated products.
- *Plant systems* are a growing option for the expression of foreign CPs. Several plant virus–derived vectors have been constructed for this purpose, the major advantages being easy plant infiltration, high replication levels, high protein yield from systemic infection, scalability, and possibility of PTMs. As with any system, plant viral expression systems have their limitations such as restrictions in insert size, inhibition of replication, and movement of viral RNA caused by the insert, biocontamination. However, some vectors that overcome these limitations of replicating viruses have been constructed for large-scale plant expression, for example, an expression vector based on cowpea mosaic virus (CPMV) sequences capable of enhancing translation [112].
- *Yeast systems* can produce glycosylated proteins but are unable to perform mammalian-type glycosylations. VLPs of some viruses such as norovirus have been produced in yeast [113].

Hepatitis B VLPs have also been produced in *Pichia pastoris* in high yield with assembly occurring downstream [114].

- *Mammalian systems* can be used to produce complex proteins having various types of modifications. They have been used, for example, for the production of vaccines but can be very costly, and these systems are prone for contamination [115, 116].
- *Insect-based systems* such as baculovirus expression systems have been used because of high protein yield, easy culturing with VLPs recovered from cultures, and their possession of eukaryotic-type PTMs. Some examples of VLPs produced using this system include influenza A virus [117, 118], hepatitis B and C viruses [119], and many viruses used as immune genes [120]. The main disadvantage is that they can easily be contaminated by baculovirus particles [121].
- *Cell-free systems* have been mostly used under in vitro conditions. This system takes advantage of the natural self-assembly abilities of the protein subunits. For example, virosomes [122] and nano-VLPs [123] have been produced using this system. MS2 bacteriophage and hepatitis B VLPs have been produced using *Escherichia coli*–based cell-free systems [124] as well as bacteriophage $\varphi\beta$-containing cytotoxic A2 protein [125].

1.5 Various Virion Structures and Properties for ENC Applications

An understanding of the high-resolution atomic structures of all components within a particle provides a rational basis for the design of nanoparticles aimed for ENC applications. The structures of several viruses have been solved mainly by X-ray crystallography and cryoelectron microscopy (cryo-EM). After the first structures of TBSV and SBMV were solved in the 1970s there have been many improvements in X-ray generators, detectors, and data processing. The requirement for X-ray analysis is to obtain well-diffracting, good-quality crystals from virions. The limitation of this technique is that no structural information on nonsymmetrical components can be obtained. The strength of cryo-EM is that lots of pictures can be taken in transmission electron microscopy (TEM) while the virions have been cooled to a metastable "vitreous" state to preserve the structure as much as possible. Similar to X-ray crystallography, cryo-EM can also be used to solve the 3D structure of a virus on atomic resolution. An even more detailed understanding of the molecular interactions within a particle can be obtained with a combination of these two techniques [126]. Techniques providing valuable structural information also include mass spectrometry (MS), nuclear magnetic resonance (NMR), small-angle X-ray scattering (SAXS), small-angle neutron scattering (SANS), dynamic light scattering (DLS), atomic force microscopy (AFM), and hydrogen deuterium exchange and mass spectrometry (HDEX).

1.5.1 Icosahedral Viruses

A systematic comparison of high-resolution structures of different virus species belonging to various families has shown that many icosahedral viruses can be placed on the basis of their MCP folds into four structural classes named picorna-, BTV-, PRD/adeno-, and HK97-like lineages [16]. Some resolved MCP structures of icosahedral viruses cannot be placed into any of these lineages. Among those are, for example, structures of bacteriophages Qβ and MS2 (ssRNA(−) genome) belonging to *Leviviridae*. They have a distinct α/β fold [127–129]. In spite of the similar MCP fold of Qβ and MS2, there are substantial differences in the protein–protein interactions within these particles. The Qβ particle structure is very stable due to tight dimer interactions and the 180 disulphide bonds, which link the dimers to the capsid structure [17, 130, 131] (Fig. 1.7). Other examples of robust viruses that contain disulfide bonds in their capsids and VLPs include HIV (genus *Lentivirus*; [132]), hepatitis B virus (HBV; family *Herpesviridae*; [133]), hepatitis C virus (HCV; family *Flaviviridae*; [134]), and John Cunningham polyomavirus (JCV; family *Polyomaviridae*; [135]). When Qβ and HBV VLPs were addressed chemically with azides and alkynes coupled to genetically engineered amino acids suitable for conjugation [136], HBV VLPs were decomposed, whereas Qβ VLPs were found to be stable, showing the need for screening different viruses for VLP stability, enabling the intended application.

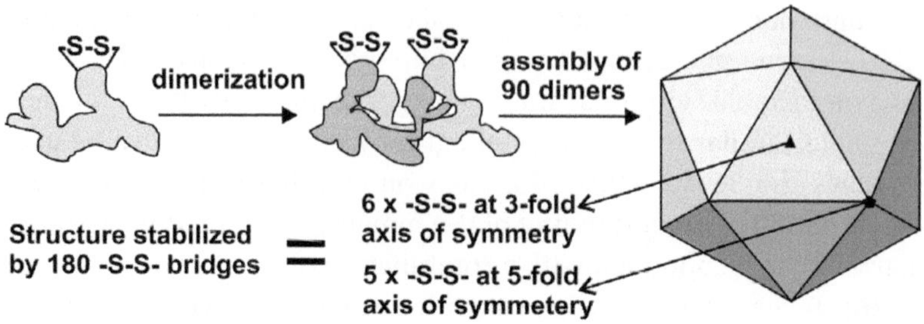

Figure 1.7 Stabilization of particles by disulphide bridges. Qβ is an example of a virion that is stabilized by 180 disulfide bridges located around the three- and five-fold axis of symmetry.

Many nonenveloped icosahedral viruses with both DNA (*Papillomaviridae, Polyomaviridae,* and *Parvoviridae*) and RNA (e.g., *Bromoviridae, Comoviridae,* and *Picornaviridae*) genomes can be placed in the picornavirus-like MCP fold group [16]. This fold is an eight-stranded antiparallel β-barrel ("jelly roll"), where two β-sheets face each other in such a way that the end strands of the four-stranded sheets form a barrel structure [137, 138]. For example, the MCPs of thoroughly studied model viruses, BMV and CCMV from the genus *Bromovirus*, have the canonical β-barrel topology with a highly basic and disordered *N*-terminal extension [139]. Research on BMV and CCMV capsids suggests that the particle structure is mainly maintained by protein–RNA interactions, which is consistent with the positively charged interior surface, whereas the protein subunit interactions are weak. An interesting property of CCMV capsids is reversible pH-dependent assembly/disassembly in the absence of viral RNA. At pH 7.5 they are dissociated into 90 separate dimers, whereas reassembly takes place at pH 5. Assembled VLPs when placed at pH higher than 6.5 become swollen, which leads to opening of 60 pores with a diameter of 2 nm, and when taken back to pH 5 the pores become closed [140, 141]. These properties have been shown to allow efficient encapsulation of an enzyme molecule and diffusion of the substrate molecule in and product out from these nanoscale reaction vessels (Fig. 1.8; [142]). When the CCMV VLP encapsidation technology was further developed, it was demonstrated that even 15 protein molecules can be efficiently encapsulated into the CCMV interior [143].

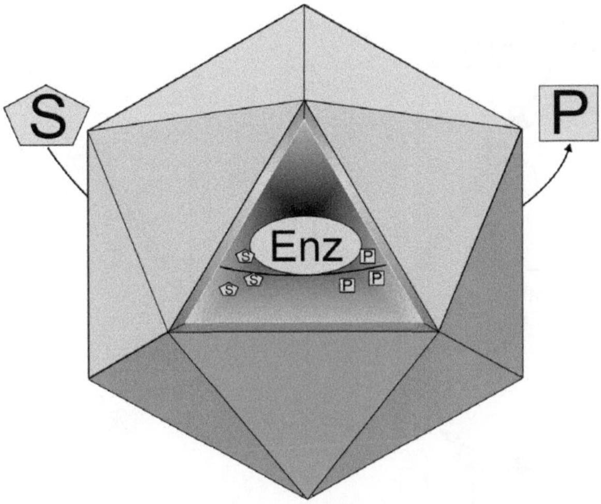

Figure 1.8 Virus particles as protein cages. Under suitable pH conditions CCMV particles can be packed with enzyme molecules. By further adjusting the pH conditions these enzyme-carrying particles can be made permeable for the substrate and product molecules, which allows the use of CCMV particles as nanoreactors. S = substrate, Enz = enzyme, and P = product.

MCPs of nine different virus families plus some nonclassified viruses share the double jelly-roll fold typical for PRD1-/adeno-like viruses. In this structure, also called the double β-barrel fold, the two β-barrels are separated by a linker region. While the MCPs with the single jelly-roll fold are typical for ssRNA viruses and only utilized by few DNA viruses, MCPs with a double jelly-roll fold are so far found only from dsDNA viruses [144]. PRD1-/adeno-type particles are complicated structures consisting of many different proteins. For example, the bacteriophage PRD1 particle consists of about 18 distinct proteins [145], and the adenovirus particle infectious for vertebrates, 12 [146] distinct proteins. More complicated particle structures provide more choice in designing novel nanoparticle types (Fig. 1.9). The proteins of adenoviral capsids, which have been genetically engineered to carry either peptide or protein modification allowing conjugation to antibodies or granting binding to an alternative cellular receptor, are the hexon, the penton, the minor capsid-stabilizing protein pIX, and the fiber knob protein [146].

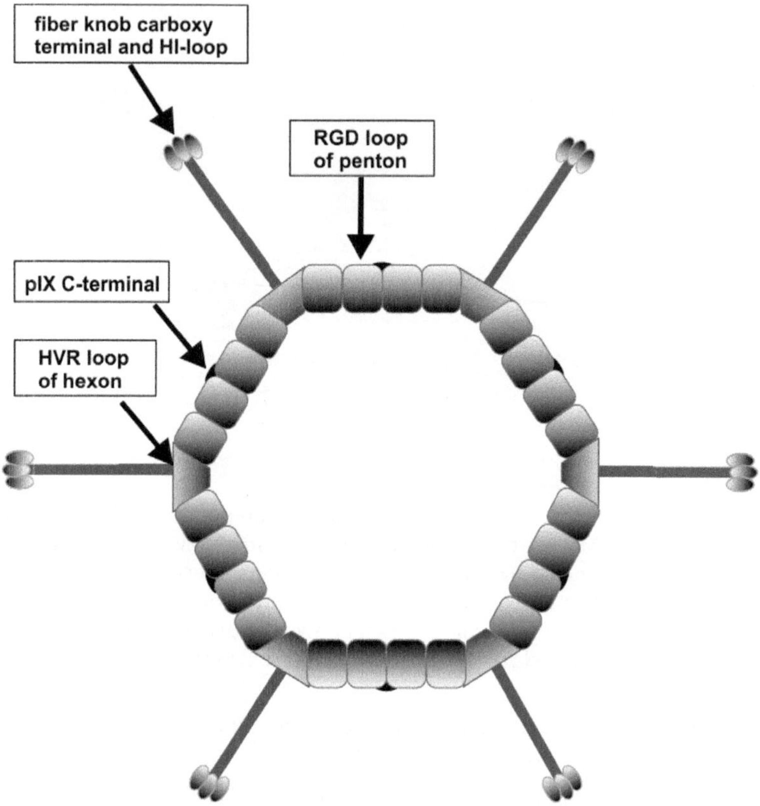

Figure 1.9 More complicated particle structures provide more choice in designing novel nanoparticle types. Four distinct structures on the adenovirus virion surface have been utilized to conjugate peptide or protein modifications to these particles.

Adenoviral capsids carrying the desired peptide and protein modifications as fusions to viral proteins can be produced from modified adenoviral vectors and targeted directly to tumor cells in vivo.

The MCP fold of siphovirus HK97 is the model for the third main fold type [147], and it consists of an axial domain having an $\alpha\beta$-structure and a peripheral domain having an extended two-strand loop at one end and a loop connecting two strands at the other end. This MCP fold is utilized by head–tail bacteriophages belonging to *Myoviridae, Siphoviridae,* and *Podoviridaeas,* as well as some dsDNA viruses (*Herpesviridae*) infecting fish, vertebrates, and molluscs [15]. An icosahedral bacteriophage P22 [148] from the family *Podoviridae*, with this type of MCP fold, has been exploited as ENCs. The P22 phage is assembled from 420 capsomers with the assistance of 100–330 copies of a scaffold protein (SP) [149]. Expression of a fluorescent protein in fusion with the SP in a heterologous expression system, also providing P22 MCP, resulted in encapsulation of many copies of SP-fluorescent protein fusions [148]. Further characterization of these ENCs showed that they have the same temperature stability as the wild-type P22 capsids, and the same morphological changes can be induced by heat, which in this system allowed enzymatic removal of the SP, while the fluorescent protein was still retained within the formed protein cages.

dsRNA viruses infecting hosts ranging from bacteria to humans are currently classified into nine families. The blue tongue virus (BTV)-like MCP fold is typical for many of these segmented and non-segmented dsRNA viruses. Bacteriophage φ6 is the type member of the family *Cystoviridae*. It shares a unique quaternary arrangement of 120 copies of the inner CP with the other members of the BTV lineage [150]. The bacteriophage φ6 PC (Fig. 1.4) has been used as a nanodevice in an in vivo RNA replication system. In this system the PC is assembled inside *E. coli* cells from expressed φ6 proteins. Within this complex it is possible to produce heterologous dsRNA molecules of up to 4.0 kb [151]. This system can potentially be used as a cost-effective way to produce dsRNA required to produce large quantities of small interfering RNAs (siRNAs) for RNA interference (RNAi) applications.

1.5.2 Helical Viruses

Rod-shaped and filamentous viruses have also found applications as ENCs. Due to the small size of the central channel, few nanometers in width, viruses with helical symmetry are not used for encapsulating enzymes. However, their structure contains more CPs per particle and a closer proximity between subunits. The majority of helical viruses are bacteriophages or viruses infecting plants, with the rare exception of the filoviruses infecting humans. MCP folds typically found from the capsids of helical viruses belong to the four-helix-bundle type.

Viruses belonging to eight genera in the family *Vigaviridae* having ssRNA(+) genomes and varicosaviruses with dsRNA genomes form rod-shaped filamentous plant viruses [152]. Tobamo-, tobra-, and furoviruses from the *Vigaviridae* family use the four-helix-bundle CP to assemble their rod-shaped virions (reviewed in [153]). The MCP fold of the flexible filaments of *Lipothrixiviridae* and stiff rod-shaped virions of *Rudiviridae*, both being dsDNA viruses infecting archaea, contain a helical bundle [126, 154]. Some similarities can be identified in the helical bundles of TMV and the archaeal viruses [16]. Many rigid rod-shaped viruses are easy to purify in high quantities, modifiable, and physically and chemically stable. This holds true especially for TMV, which accumulates to levels of up to 10 g/kg of leaf biomass. The externally located *N*-terminus of the TMV CP has been modified with a reactive lysine, allowing biotinylation of the capsid [155]. Subsequently, a streptavidin fusion peptide could be linked to TMV particles, and this strategy was used to create TMV–antigen complexes that were found to be more immunogenic than the uncoupled antigen alone. One of the problems related to rod-shaped virions are the steric problems caused by the close proximity of the CP subunits and consequently decorating proteins to each other when the enzyme and the CP are genetically fused. A possible solution to this problem is to engineer a long enough linker between the enzyme and the CP to allow more space for the enzyme (Fig. 1.10). A flexible linker engineered to the *C*-terminus of the TMV CP was found as a solution to fit larger peptides on the virion surface [156].

Flexible filamentous plant viruses belong to 20 genera and 3 families (*Potyviridae, Flexiviridae*, and *Closteroviridae*). This particle structure has been evolutionary successful in plants, as it is conserved among so many plant viruses. Similarities between potexvirus (family *Flexiviridae*) and potyvirus (family *Potyviridae*) symmetry, in the low-resolution structural models and high-resolution fiber diffraction data, show that the MCPs of these flexible filamentous viruses are structurally related [157]. The flexibility of these viruses could be explained by a higher degree of freedom in the CP–CP interactions due to a smaller contact area than in the case of rigid rods. Due to this difference flexible viruses have more variation in their symmetry parameters, smaller diameters, and deeper grooves in their structure than rigid rods [158]. Potato virus X (PVX) has been decorated with a lipase enzyme. To avoid the steric hindrance problem the enzyme was genetically fused to CPs via 2A peptide of the food and mouth disease virus [159]. The 2A peptide has the property of self-cleavage, which, however, happens only to a certain percentage of the fusion molecules. This strategy leads to a mixture of CPs and enzyme–CP building blocks for the assembly of ENCs, as illustrated in Fig. 1.10.

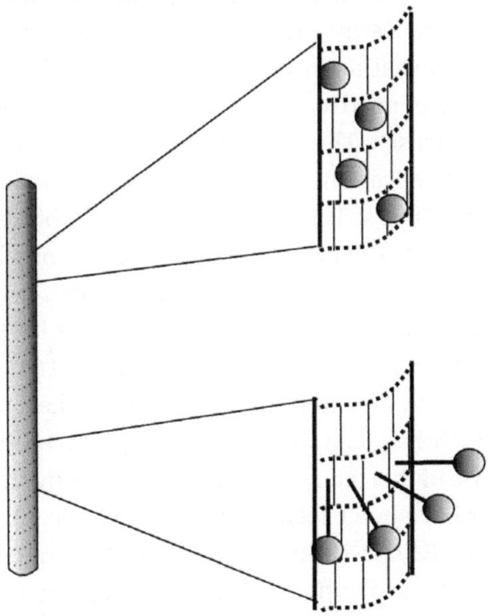

Figure 1.10 Filamentous virus particles as ENCs. Illustrated are two strategies to avoid steric problems in packaging the enzyme–CP fusion into the filamentous particles. The ENCs may contain CPs and enzyme–CP fusions in a ratio, allowing efficient packaging or a long enough linker to be placed between the enzyme and the CP to give space for the enzyme.

The filamentous bacteriophages belong to the *Inovirus* and *Pectrovirus* genera. Bacteriophages fd, f1, and M13 belonging to inoviruses have become extremely useful vehicles for peptide display, a technique to present peptides on the virion surface. This methodology has been instrumental for many interaction studies. The desired peptide can be fused either to the minor CP pIII present in five copies per particle or to the MCP pVIII. Phage display technology is a versatile tool used, for example, in antibody production and enzyme technology [160]. As an instrument to advance enzyme nanocarrier technology, it can be used to screen for attachment petides capable of binding to virus particles.

1.5.3 Enveloped Viruses

The assembly pathway of many viruses incorporates an envelope that, depending on the virus group, originates from different kinds of cellular membranes into the capsid organization. The envelope can be either external as in retroviruses and influenza viruses or located internally beneath an outer proteinaceous capsid layer, as in alpha- and flaviviruses. The majority of viral proteins embedded in envelopes are glycoproteins, which exist there in the form of either homo- or heterodimers. Their functions include binding to specific receptors at the host cell surface and induction of membrane fusions. The membrane fusion property of these viruses may become applicable in fabricating composite materials, as lipid bilayers can be assembled on several kinds of solid supports [161].

The viruses with an external envelope often lack symmetric organization, which has complicated their high-resolution structural analysis. In spite of this many structures of the individual viral membrane proteins have been solved, for example, recently the alphaviral glycoprotein E1/E2 complex [162, 163]. Surface engineering of enveloped viruses is widely used for vector generation [164]. Although some difficulties have been faced in retaining the receptor binding and fusion properties of the envelope proteins, these modifications interfere less with self-assembly of the virus than genetic engineering of the CPs. The membrane fusion proteins from baculovirus [165, 166], avian leucosis virus [167], vesicular stomatitis virus [168], murine leukemia virus (MLV; [169]), and influenza virus [170] have permissive insertion sites for foreign peptide sequences. The size of the insertion may lead to diminished infectivity, but, for example, structurally complex mammalian enzymes, such as tissue plasminogen activator, can be displayed on the surface of MLV [171]. This technology can be used, for example, as an eukaryotic virus display system that, as an expansion to the phage display technology, allows screening structurally complex proteins requiring disulphide bridging, glycosylation, and/or membrane anchorage.

1.6 New Viral Species to Be Exploited as Viral ENCs

The number of viral orders has increased from 1 to 6 and families from 50 to 87 during the past 17 years (see Fig. 1.2). New insights into virus diversity and abundance of different virus types in a variety of environments are today gained from metagenomics studies. As an example, distribution of plant virus families in wild plants differs remarkably from the distribution map based on plant viruses currently recognized by the International Committee of Taxonomy of Viruses (ICTV) [172]. New virus species have been discovered along the growing interest toward the domain Archaea [5, 173, 174]. Approximately 100 viruses infecting archaea have been isolated and characterized until now in contrast to over 6000 characterized bacterial viruses [5, 175, 176]. Archaea are numerous in the oceans and in a broad range of other habitats. Many of them are also extremophiles living in harsh conditions, and therefore viruses that infect extremophiles have adapted to such conditions as well. Many new virion architectures have been observed in archaeal viruses. Acidianus two-tailed virus (ATV) belonging to Bicaudaviridae is a peculiar virus. When outside the host cell, this spindle-shaped virus grows two tails extending from the poles of the spindle [177]. ATV MCP is mainly alpha-helical, but the fold differs from that of the four-helix-bundle capsids and forms its own group of MCP folds. Aeropyrum coil-shaped virus (ACV) of the hyperthermophilic archaeon *Aeropyrumpernix* has a novel nonenveloped, hollow, cylindrical virion architecture [178]. In this structure two intertwining halves of a single circular nucleoprotein form a coiling fiber. Halorubrum pleomorphic virus 1 (HRPV-1) is a pleolipovirus, which means that its genome is enclosed in a lipid vesicle without associated nucleoproteins [179]. Two major structural protein species are associated with HRPV-1 virions, one forming spikes randomly distributed on the virion surface and the other embedded in the inner surface of the vesicle membrane. The diversity of novel virion morphologies found in archaeal viruses extends much further than the few examples described here. These viruses may provide a pool for novel properties beneficial for the ENC technology.

Acknowledgments

The authors thank Dr. Daniela Cardinale for drawing Figs. 1.3 and 1.6 and Table 1.1. Dr. Maija Pietilä is thanked for Figs. 1.1 and 1.2. The Academy of Finland is acknowledged for financial support (decision nos. 134684 and 138329 to K. M.).

References

1. Culver, J. N. (2002) Tobacco mosaic virus assembly and disassembly: Determinants in pathogenicity and resistance, *Ann. Rev. Phytopathol.*, **40**, 287–308.
2. Ivanov, K. I., and Mäkinen, K. (2012) Coat proteins, host factors and plant viral replication, *Curr. Opin. Virol.*, **6**, 712–718.
3. Klug, A. (1999) The tobacco mosaic virusparticle: structure and assembly, *Phil. Trans. R. Soc. Lond. B*, **354**, 531–535.
4. Crick, F. H. C., and Watson, J. D. (1956) Structure of small viruses, *Nature*, **177**, 473–475.
5. Pina, M., Bize, A., Forterre, P., and Prangishvili, D. (2011) The archeoviruses, *FEMS Microbiol. Rev.*, **35**, 1035–1054.
6. King, A. M. Q., Adams, M. J., Carstens, E. B., and Lefkowitz, E. J. (2012) *Virus Taxonomy: Ninth Report of the International Committee on Taxonomy of Viruses*, Vol. 9 (Elsevier Academic, San Diego, CA).
7. Suttle, C. A. (2007) Marine viruses: major players in the global ecosystem, *Nat. Rev. Microbiol.*, **5**, 801–812.
8. Harrison, S. C., Olson, A. J., Schutt, C. E., Winkler, F. I., and Bricogne, G. (1978) Tomato bushy stunt virus at 2.9Å resolution, *Nature*, **276**, 368–373.
9. Abad-Zapatero, C., Abdel-Meguid, S. S., Johanson, J. E., Leslie, A. G. W., Raymet, I., Rossmann, M. G., Suck, D., and Tsukihara T. (1980) Structure of Southern bean mosaic virus at 2.8Å resolution, *Nature*, **286**, 33–39.
10. Rossmann, M. G., Arnold, E., Erickson, J. W., Frankenberger, E. A., Griffith, J. P., Hecht, H. J., Johnson, J. E., Kamer, G., Luo., M., Mosser, A. G., Rueckert, R. R., Sherry, B., and Vriend, G. (1985) Structure of a human common cold virus and functional relationship to other picornaviruses, *Nature*, **317**, 145–153.
11. Hogle, J. M., Chow, M., and Filman, D. J. (1985) Three-dimensional structure of poliovirus at 2.9Å resolution, *Science*, **229**, 1358–1365.

12. Bamford, D. H., Burnett, R. M., and Stuart, D. I. (2002) Evolution of viral structure, *Theor. Popul. Biol.*, **61**, 461–470.

13. Benson, S. D., Bamford, J. K., Bamford, D. H., and Burnett, R. M. (2004) Does a common architecture reveal a viral lineage spanning all three domains of life? *Mol. Cell,* **16**, 673–685.

14. Bamford, D. H., Grimes, J. M., and Stuart, D. I. (2005) What does structure tell us about virus evolution? *Curr. Opin. Struct. Biol.,* **15**, 655–663.

15. Krupovic, M., and Bamford, D. H. (2011) Double-stranded DNA viruses: 20 families and only five different architectural principles for virion assembly, *Curr. Opin. Virol.*, **1**, 118–124.

16. Abrescia, N. G. A., Bamford, D. H., Grimes, J. M., and Stuart D. I. (2012) Structure unifies the viral universe, *Annu. Rev. Biochem.*, **81**, 795–822.

17. Ashcroft, A. E., Lago, H., Macedo, J. M. B., Horn, W. T., Stonehouse, N. J., and Stockley, P. G. J. (2005) Engineering thermal stability in RNA phage capsids via disulphide bonds, *Nanosci. Nanotechnol.*, **5**, 2034–2041.

18. Caldeira, J. C., and Peabody, D. S. J. (2011) Thermal stability of RNA phage virus-like particles displaying foreign peptides, *Nanobiotechnology,* **9**, 22–28.

19. Johnson, H. R., Hooker, J. M., Francis, M. B., and Clark, D. S. (2007) Solubilization and stabilization of bacteriophage MS2 in organic solvents, *Biotechnol. Bioeng.*, **97**, 224–234.

20. Koudelka, K. J., and Manchester, M. (2010) Chemically modified viruses: principles and applications, *Curr. Opin. Chem. Biol.,* **14**, 810–817.

21. Fraenkel-Conrat, H., and Williams, R. C. (1955) Reconstitution of active tobacco mosaic virus from its inactive protein and nucleic acid components, *Proc. Natl. Acad. Sci. U. S. A.*, **41**, 690–698.

22. Caspar, D. L. D., and Klug, A. (1962) Physical principles in the construction of regular viruses, *Cold Spring Harbor Symp.*, **27**, 1–24.

23. Fane, B. A., and Prevelige, P. E. J. (2003) Mechanism of scaffolding-assisted virus assembly, in *Virus Structure*, Vol. 64, eds. Chiu, W., and Johnson, J. E. (Academic Press, San Diego, CA), 259–299.

24. Annamalai, P., and Rao A. L. (2005) Dispensability of 3' tRNA-like sequence for packaging cowpea chlorotic mottle virus genomic RNAs, *Virology,* **332**, 650–658.

25. Annamalai, P., and Rao, A. L. (2005) Replication-independent expression of genome components and capsid protein of brome mosaic virus *in planta*: a functional role for viral replicase in RNA packaging, *Virology,* **338**, 96–111.

26. Qu, F., and Morris, T. J. (1997) Encapsidation of turnip crinkle virus is defined by a specific packaging signal and RNA size, *J. Virol.*, **71**, 1428–1435.

27. Kelloniemi, J., Mäkinen, K., and Valkonen, J. P. (2008) Three heterologous proteins simultaneously expressed from a chimeric potyvirus: infectivity, stability and the correlation of genome and virion lengths, *Virus Res.*, **135**: 282–291.

28. Caspar, D. L. D. (1963) Assembly and stability of the tobacco mosaic virus particle, *Adv. Prot. Chem.*, **18**, 37–121.

29. Durham, A. C., Finch, J. T., and Klug, A. (1971) States of aggregation of tobacco mosaic virus protein, *Nat. New Biol.*, **229**, 37–42.

30. Durham, A. C., and Klug, A. (1971) Polymerization of tobacco mosaic virus protein and its control, *Nat. New Biol.*, **229**, 42–46.

31. Butler, P. J., Klug, A. 1971. Assembly of the particle of tobacco mosaic virus from RNA and disks of protein, *Nat New Biol.* **229**, 47–50.

32. Zimmern, D., and Butler, P. J., (1977) The isolation of tobacco mosaic virus RNA fragments containing the origin for viral assembly, *Cell*, **11**, 455–462.

33. Zimmern, D. (1977) The nucleotide sequence at the origin for assembly on tobacco mosaic virus RNA, *Cell*, **11**, 463–482.

34. Jonard, G., Richards, K. E., Guilley, H., and Hirth, L. (1977) Sequence from the assembly nucleation region of TMV RNA, *Cell*, **11**, 483–493.

35. Sleat, D. E., Gallie, D. R., Watts, J. W., Deom, C. M., Turner, P. C., Beachy, R. N., and Wilson, T. M. (1988) Selective recovery of foreign gene transcripts as virus-like particles in TMV-infected transgenic tobaccos, *Nucleic Acids Res.*, **16**, 3127–3140.

36. Sleat, D. E., Plaskitt, K. A., and Wilson, T. M. (1988) Selective encapsidation of CAT gene transcripts in TMV-infected transgenic tobacco inhibits CAT synthesis, *Virology*, **165**, 609–612.

37. Butler, P. J., Finch, J. T., and Zimmern, D. (1977) Configuration of tobacco mosaic virus, RNA during virus assembly, *Nature*, **265**, 217–219.

38. Lebeurier, G., Nicolaieff, A., and Richards, K. E. (1977) Inside-out model for self-assembly of tobacco mosaic virus, *Proc. Natl. Acad. Sci. U. S. A.*, **74**, 149–153.

39. Junker-Niepmann, M., Bartenschlager, R., and Schaller, H. (1990) A short cis-acting sequence is required for hepatitis B virus pregenome encapsidation and sufficient for packaging of foreign RNA, *EMBO J.*, **9**, 3389–3396.

40. Choi, Y. G., Dreher, T. W., and Rao, A. L. (2002) tRNA elements mediate the assembly of an icosahedral RNA virus, *Proc. Natl. Acad. Sci. U. S. A.*, **99**, 655–660.

41. Frolova, E., Frolov, I., and Schlesinger, S. (1997) Packaging signals in alphaviruses, *J. Virol.*, **71**, 248–258.

42. Vriend, G., Verduin, B. J., and Hemminga, M. A. (1986) Role of the N-terminal part of the coat protein in the assembly of cowpea chlorotic mottle virus. A 500 MHz proton nuclear magnetic resonance study and structural calculations, *J. Mol. Biol.*, **191**, 453–460.

43. Hamilton, R. I., and Tremaine, J. H. (1996) Dianthoviruses: properties, molecular biology, ecology and control, in *The Plant Viruses (Vol. 5: Polyhedral Virions and Biparetite RNA Genomes)*, eds. Harrison, B. D., and Murant, A. F. (Plenum Press, New York), 251–282.

44. Rao, A. L., and Grantham, G. L. (1996) Molecular studies on bromovirus capsid protein. II. Functional analysis of the amino-terminal arginine-rich motif and its role in encapsidation, movement, and pathology, *Virology*, **226**, 294–305.

45. Sacher, R., and Ahlquist, P. (1989) Effects of deletions in the N-terminal basic arm of brome mosaic virus coat protein on RNA packaging and systemic infection, *J. Virol.*, **63**, 4545–4552.

46. Linger, B. R., Kunovska, L., Kuhn, R. J., and Golden, B. L. (2004) Sindbis virus nucleocapsid assembly: RNA folding promotes capsid protein dimerization, *RNA*, **10**, 128–138.

47. Karran, R., and Hudak, K. (2011) Depurination of Brome mosaic virus RNA3 inhibits its packaging into virus particles, *Nucleic Acids Res.*, **39**, 7209–7222.

48. Lomonossoff, G. P., and Butler, P. J. (1979) Location and encapsidation of the coat protein cistron of tobacco mosaic virus. A bidirectional elongation of the nucleoprotein rod, *Eur. J. Biochem.*, **93**, 157–164.

49. Fairall, L., Finch, J. T., Hui, C. F., Cantor, C. R., and Butler, P. J. (1986) Studies of tobacco mosaic virus reassembly with an RNA tail blocked by a hybridised and cross-linked probe, *Eur. J. Biochem.*, **156**, 459–465.

50. Turner, D. R., McGuigan, C. J., and Butler, P. J. G. (1989) Assembly of hybrid RNAs with tobacco mosaic virus coat protein. Evidence for incorporation of disks in 5' elongation along the major RNA tail, *J. Mol. Biol.*, **209**, 407–422.

51. Butler, P. J. G. (1984) The current picture of the structure and assembly of tobacco mosaic virus, *J. Gen. Virol.*, **65**, 253–279.

52. Bazinet, C., and King J. (1985) The DNA translocating vertex of dsDNA bacteriophage, *Annu. Rev. Microbiol.*, **39**, 109–129.

53. Newcomb, W. W., Juhas, R. M., Thomsen, D. R., Homa, F. L., Burch, A. D., Weller, S. K., and Brown, J. C. (2001) The UL6 gene product forms the portal for entry of DNA into the herpes simplex virus capsid, *J. Virol.*, **75**, 10923–10932.

54. Poranen, M., and Tuma, R. (2004) Self-assembly of double-stranded RNA bacteriophages, *Virus Res.*, **101**, 93–100.

55. Poranen, M. M., Paatero, A. O., Tuma, R., and Bamford, D. H. (2001) Self-assembly of a viral molecular machine from purified protein and RNA constituents, *Mol. Cell*, **7**, 845–854.

56. Tilsner, J., and Oparka, K. J. (2012) Missing links? The connection between replication and movement of plant RNA viruses, *Curr. Opin. Virol.*, **2**, 699–705.

57. Den Boon, J. A., Diaz, A., and Ahlquist, P. (2010) Cytoplasmic viral replication complexes, *Cell Host Microbe*, **8**, 77–85.

58. Laliberté, J.-F., and Sanfaçon, H. (2010) Cellular remodeling during plant virus infection, *Annu. Rev. Phytopathol.*, **48**, 69–91.

59. Verchot, J. (2011) Wrapping membranes around plant virus infection, *Curr. Opin. Virol.*, **1**, 388–395.

60. Tilsner, J., Linnik, O., Wright, K. M., Bell, K., Roberts, A. G., Lacomme, C., Santa Cruz, S., and Oparka, K. J. (2012) The TGB1 movement protein of potato virus X reorganizes actin and endomembranes into the X-body, a viral replication factory, *Plant Physiol.*, **158**, 1359–1370.

61. Patarroyo, C., Laliberté, J-F., and Zheng, H. (2013) Hijack it, change it: how do plant viruses utilize the host secretory pathway for efficient viral replication and spread? *Front. Plant Sci.*, **3**, 1–8.

62. Kao, C. C., Ni, P., Hema, M., Huang, X., and Dragnea, B. (2011) The coat leads the way: an update on basic and applied studies with Brome mosaic virus coat protein, *Mol. Plant Path.*, **12**, 403–412.

63. Martinière, A., Bak, A., Macia, J. L., Lautredou, N., Gargani, D., Doumayrou, J., Garzo, E., Moreno, A., Fereres, A., Blanc, S., and Drucker, M. (2013) A virus responds instantly to the presence of the vector on the host and forms transmission morphs, *Elife*, **2**, e00183.

64. Stubbs, G. (1999) Tobacco mosaic virus particle structure and the initiation of assembly, *Phylos. Trans. R. Soc. Lond. B*, **354**, 551–557.

65. Shaw, J. G. (1999) Tobacco mosaic virus and the study of early events in virus infection, *Phylos. Trans. R. Soc. Lond. B*, **354**, 603–611.

66. Hull, R. (2009) Expression of viral genomes, in *Comparative Plant Virology, Second Edition*, Amsterdam; Boston: Elselvier Academic press, 119–121.

67. Kurtz-Fritsch, C., and Hirth, L. (1972) Uncoating of two spherical plant virus, *Virology*, **47**, 385–396.

68. Kasamatsu, H., and Nakanishi, A. (1998) How do animal DNA viruses get the nucleus? *Annu. Rev. Microbiol.*, **52**, 627–686.

69. Kobiler, O., Drayman, N., Butin-Israeli, V., and Oppenheim, A. (2012) Virus strategies for passing the nuclear envelope barrier, *Nucleus*, **3**, 526–539.

70. Hulo, C., de Castro, E., Masson, P., Bougueleret, L., Bairoch, A., Xenarios, I., and Le Mercier, P. (2011) ViralZone: a knowledge resource to understand virus diversity, *Nucleic Acids Res.*, **39**, D576–D582.

71. Masson, P., Hulo, C., De Castro, E., Bitter, H., Gruenbaum, L., Essioux, L., Bougueleret, L., Xenarios, I., and Le Mercier, P. (2013) Viralzone: recent updates to the virus knowledge resource, *Nucleic Acids Res.*, **41**, D579–D583.

72. Greber, U. F., Singh, I., and Helenius, A. (1994) Mechanisms of virus uncoating, *Trends Microbiol.*, **2**, 52–56.

73. Haywood, A. M. (2010) Membrane uncoating of intact enveloped viruses, *J. Virol.*, **84**, 10946–10955.

74. Rao, A. L. N. (2006) Genome packaging by spherical plant RNA viruses, *Annu. Rev. Phytopathol.*, **44**, 61–87.

75. Shneemann, A. (2006) The structural and functional role of RNA in icosahedral virus assembly, *Annu. Rev. Microbiol.*, **60**, 51–67.

76. Sun, S., Rao, V. B., and Rossmann, M. G. (2010) Genome packaging in viruses, *Curr. Opin. Struct. Biol.*, **20**, 114–120.

77. Jewell, N. A., and Mansky, L. M. (2000) In the beginning: genome recognition, RNA encapsidation and the initiation of complex retrovirus assembly, *J. Gen. Virol.*, **81**, 1889–1899.

78. Choi, Y. G., and Rao, A. L. (2003) Packaging of brome mosaic virus RNA3 is mediated through a bipartite signal, *J. Virol.*, **77**, 9750–9757.

79. Sullivan, M. I., and Ahlquist, P. (1999) A brome mosaic virus intergenic RNA3 replication signal functions with viral replication protein 1a to dramatically stabilize RNA in vivo, *J. Virol.*, **73**, 2622–2632.

80. Annamalai, P., Rofail, F., DeMason, D. A., and Rao, A. L. N. (2008) Replication-coupled packaging mechanism in positive-strand RNA viruses: synchronized coexpression of functional multigenome RNA components of an animal and a plant virus in *Nicotiana benthamiana* cells by agroinfiltration, *J. Virol.*, **82**, 1484–1495.

81. Seo, J.-K., Kwon, S.-J., and Rao, L. N. (2012) Physical interaction between viral replicase and capsid protein is required for genome-packaging specificity in an RNA virus, *J. Virol.*, **86**, 6210–6221.

82. Nugent, C. I., Johnson, K. L., Sarnow, P., and Kirkegaard, K. (1999) Functional coupling between replication and packaging of poliovirus replicon RNA, *J. Virol.*, **73**, 427–435.

83. Venter, P. A., Krishna, N. K., and Schneemann, A. (2005) Capsid protein synthesis from replicating RNA directs specific packaging of the genome of a multipartite, positive-strand RNA virus, *J. Virol.*, **79**, 6239–6248.

84. Khromyhk, A. A., Varnavsky, A. N., Sedlak, P. L., and Westaway, E. G. (2001) Coupling between replication and packaging of flavivirus RNA: evidence derived from the use of DNA-based full-length cDNA of Kunjin virus, *J. Virol.*, **75**, 4633–4640.

85. Satyanarayana, T., Gowda, S., Ayllón, M. A., and Dawson, W. O. (2004) Closterovirus bipolar virion: evidence for initiation of assembly by minor coat protein and its restriction to the genomic RNA 5' region, *Proc. Natl. Acad. Sci. U. S. A*, **101**, 799–804.

86. Torrance, L., Andreev, I. A., Gabrenaite-Verkhovskaya, R., Cowan, G., Mäkinen, K., and Taliansky, M. E. (2006) An unusual structure at one end of potato potyvirus particles, *J. Mol. Biol.*, **357**, 1–8.

87. Durham, A. C., and Klug, A. (1972) Structures and roles of the polymorphic forms of tobacco mosaic virus protein. 3. A model for the association of A-protein into disks, *J. Mol. Biol.*, **67**, 315–332.

88. Butler, P.J., and Klug, A. (1971) Assembly of the particle of tobacco mosaic virus from RNA and disks of protein, *Nat. New Biol.*, **229**, 47–50.

89. Okada, Y., and Ohno, T. (1972) Assembly mechanism of tobacco mosaic virus from its ribonucleic acid and protein, *Mol. Gen. Genet.*, **114**, 205–213.

90. Lauffer, M. A., and Stevens, C. L. (1968) Structure of the tobacco mosaic virus particle; polymerization of tobacco mosaic virus protein, *Adv. Virus Res.*, **13**, 1–63.

91. Culver, J. N., Stubbs, G., and Dawson, W. O. (1994) Structure-function relationship between tobacco mosaic virus coat protein and hypersensitivity in *Nicotiana sylvestris*, *J. Mol. Biol.*, **242**, 130–138.

92. Mandelkow, E., Stubbs, G., and Warren, S. (1981) Structures of the helical aggregates of tobacco mosaic virus protein, *J. Mol. Biol.*, **152**, 375–386.

93. Incardona, N. L., and Kaesberg, P. (1964) A Ph-induced structural change in brome grass mosaic virus, *Biophys. J.*, **4**, 11–21.

94. Bancroft, J. B., Hills, G. J., and Markham, R. (1967) A study of the self-assembly process in a small spherical virus. Formation of organized structures from protein subunits in vitro, *Virology*, **31**, 354–379.

95. Bancroft, J. B., Wagner, G. W., and Bracker, C. E. (1968) The self-assembly of a nucleic-acid free pseudo-top component for a small spherical virus, *Virology* **36**, 146–149.
96. Hiebert, E., and Bancroft, J. B. (1969) Factors affecting the assembly of some spherical viruses, *Virology*, **39**, 296–311.
97. Brakke, M. K. (1963) Stabilization of brome mosaic virus by magnesium and calcium, *Virology*, **19**, 367–374.
98. Sorger, P. K., Stockley, P. G., and Harrison, S. C. (1986) Structure and assembly of turnip crinkle virus. II. Mechanism of reassembly in vitro, *J. Mol. Biol.*, **191**, 639–658.
99. Golden, J. S., and Harrison, S. C. (1982) Proteolytic dissection of turnip crinkle subunits in solution, *Biochemistry*, **21(16)**, 3862–3866.
100. Goodman, R. M. (1975) Reconstitution of potato virus X in vitro. I. Properties of the dissociated protein structural subunits, *Virology*, **68**, 287–298.
101. Goodman, R. M., McDonald, J. G., Horne, R. W., and Bancroft, J. B. (1976) Assembly of flexuous plant viruses and their proteins, *Philos. Trans. R. Soc. Lond. B*, **276**, 173–179.
102. McDonald, J. G., Beveridge, T. J., and Bancroft, J. B. (1976) Self-assembly of protein from a flexuous virus, *Virology*, **69**, 327–331.
103. Cleland, W. W. (1964) Dithiothreitol, a new protective reagent for SH groups, *Biochemistry*, **3**, 480–482.
104. Hiebert, E., Bancroft, J. B., and Bracker, C. E. (1968) The assembly in vitro of some small spherical viruses, hybrid viruses, and other nucleoproteins, *Virology*, **34**, 492–508.
105. Rolfsson, O., Toropova, K., Morton, V., Francese, S., Basnak, G., Thompson, G. S., Homans, S. W., Ashcroft, A. E., Stonehouse, N. J., Ranson, N. A., and Stockley, P. G. (2008) RNA packing specificity and folding during assembly of the bacteriophage MS2, *Comput. Math. Methods Med.*, **9**, 339–349.
106. Caldeira, J. C., and Peabody, D. S. (2007) Stability and assembly in vitro of bacteriophage PP7 virus-like particles, *J. Nanobiotech.*, **5**, 1–10.
107. Siddhartha, A., Datta, K., Temeselew, L. G., Crist, R. M., Soheilian, F., Kamata, A., Mirro, J., Harvin, D., Nagashima, K., Cachau, R. E., and Rein, A. (2011) On the role of the SP1 domain in HIV-1 particle assembly: a molecular switch? *J Virol.*, **85**, 4111–4121.
108. Hadravová, R., de Marco, A., Ulbrich, P., Štokrová, J., Doležal, M., Pichová, I., Ruml, T., Briggs, J. A. G., and Rumlová, M. (2012) *In vitro* assembly of virus-like particles of a gammaretrovirus, the murine leukemia virus XMRV, *J. Virol.*, **86**, 1297–1306.

109. Hema, M., Subba Reddy, ChV., Savithri, H. S., and Sreenivasulu, P. (2008) Assembly of recombinant coat protein of sugarcane streak mosaic virus into potyvirus-like particles, *Indian J. Exp. Biol.*, **46**, 793–796.

110. Jacob, T., and Usha, R. (2002) Expression of cardamom mosaic virus coat protein in Escherichia coli and its assembly into filamentous aggregates, *Virus Res.*, **86**, 133–141.

111. Jagadish, M. N., Hamilton, R. C., Fernandez, C. S., Schoofs, P., Davern, K. M., Kalnins, H., Ward, C. W., and Nisbet, I. T. (1993) High level production of hybrid potyvirus-like particles carrying repetitive copies of foreign antigens *in Escherichia* coli, *Nat. Biotechnol.*, **11**, 1166–1170.

112. Sainsbury, F., and Lomonossoff, G. P. (2008) Extremely high-level and rapid transient protein production in plants without the use of viral replication, *Plant Physiol.*, **148**, 1212–1218.

113. Xia, M., Farkas, T., and Jiang, X. (2007) Norovirus capsid protein expressed in yeast forms virus-like particles and stimulates systemic and mucosal immunity in mice following an oral administration of raw yeast extracts, *J. Med. Virol.*, **79**, 74–83.

114. Lünsdorf, H., Gurramkonda, C., Adnan, A., Khanna, N., and Rinas, U. (2011) Virus-like particle production with yeast: ultrastructural and immunocytochemical insights into *Pichia pastoris* producing high levels of the hepatitis B surface antigen, *Microb. Cell Fact.*, **10**, 48, doi:10.1186/14752859-10-48.

115. Grillberger, L., Kreil, T. R., Nasr, S., and Reiter, M. (2009) Emerging trends in plasma-free manufacturing of recombinant protein therapeutics expressed in mammalian cells, *Biotechnol. J.*, **4**, 186–201.

116. Walter, J. K., Werz, W., and Berthold, W. (1996)Process scale considerations in evaluation studies and scale-up, *Dev. Biol. Stand.*, **88**, 99–108.

117. Latham, T., and Galarza, J. M. (2001) Formation of wild-type and chimeric influenza virus-like particles following simultaneous expression of only four structural proteins, *J. Virol.*, **75**, 6154–6165.

118. Wu, C. Y., Yeh, Y. C., Yang, Y. C., Chou, C., Liu, M. T., Wu, H. S., Chan, J. T., Hsiao, P. W. (2010) Mammalian expression of virus-like particles for advanced mimicry of authentic influenza virus, *PLOS ONE*, **5**(3), e9784.

119. Baumert, T. F., Ito, S., Wong, D. T., and Liang, T. J. (1998) Hepatitis C virus structural proteins assemble into virus like particles in insect cells, *J. Virol.* **72**, 3827–3836.

120. Noad, R., and Roy, P. (2003) Virus-like particles as immunogens, *Trends Microbiol.*, **11**, 438–444.

121. Hervas-Stubbs, S., Rueda P., Lopez, L., and Leclerc, C. (2007) Insect baculoviruses strongly potentiate adaptive immune responses by inducing type I IFN, *J. Immunol.,* **178**, 2361–2369 [Erratum in *J. Immunol.* (2007), **178**, 6653].

122. Wiedermann, U., Wiltschke, C., Jasinska, J., Kundi, M., Zurbriggen, R., Garner-Spitzer, E., Bartsch, R., Steger, G., Pehamberger, H., Scheiner, O., and Zielinski, C. C. (2010) A virosomal formulated Her-2/neu multi-peptide vaccine induces Her-2/neu specific immune responses in patients with metastatic breast cancer: a phase I study, *Breast Cancer Res. Treat.*, **119**, 673–683.

123. Kaba, S. A., Brando, C., Guo, Q., Mittelholzer, C., Raman, S., Tropel, D., Aebi, U., Burkhard, P., and Lanar, D. E. (2009) A non-adjuvanted polypeptide nanoparticle vaccine confers long-lasting protection against rodent malaria, *J. Immunol.*, **183**, 7268–7277.

124. Bundy, B. C., Franciszkowicz, M. J., and Swartz, J. R. (2008) *Escherichia coli*-based cell-free synthesis of virus-like particles, *Biotechnol. Bioeng.*, **100**, 28–37.

125. Smith, M. T., Varner, C. T., Bush, D. B., and Bundy, B. C. (2012) The incorporation of the A2 protein to produce novel Qβ virus-like particles using cell-free protein synthesis, *Biotechnol. Prog.*, **28**, 549–555.

126. Szymczyna, B. R., Taurog, R. E., Young, M. J., Snyder, J. C., Johnson, J. E., and Williamson, J. R. (2009) Synergy of NMR, computation, and X-ray crystallography for structural biology, *Structure*, **17**, 499–507.

127. Valegård, K., Liljas, L., Fridborg, K., and Unge, T. (1990) The three-dimensional structure of the bacterial virus MS2, *Nature*, **345**, 36–41.

128. Golmohammadi, R., Valegård, K., Fridborg, K., and Liljas, L. (1993) The refined structure of bacteriophage MS2 at 2.8 Å resolution, *J. Mol. Biol.*, **234**, 620–639.

129. Golmohammadi, R., Fridborg, K., Bundule, M., Valegård, K., and Liljas, L. (1996) The crystal structure of bacteriophage Q beta at 3.5 Å resolution, *Structure*, **4**, 543–554.

130. Fiedler, J. D., Higginson, C., Hovlid, M. L., Kislukhin, A. A., Castillejos, A., Manzenrieder, F., Campbell, M. G., Voss, N. R., Potter, C. S., Carragher, B., and Finn, M. G. (2012) Engineered mutations change the structure and stability of a virus-like article, *Biomacromolecules,* **13**, 2339–2348.

131. Bundy, B. C., and Swartz, J. R. (2011) Efficient disulphide bond formation in virus-like particles, *J. Biotech.*, **154**, 230–239.

132. Ivanov, D., Tsodikov, O. V., Kasanov, J., Ellenberger, T., Wagner, G., and Collins, T. (2007) Domain-swapped dimerization of the HIV-1 capsid C-terminal domain, *Proc. Natl. Acad. Sci. U. S. A.*, **104**, 4353–4358.

133. Steven, A. C., Conway, J. F., Cheng, N., Watts, N. R., Belnap, D. M., Harris, A., Stahl, S. J., and Wingfield, P. T. (2005) Structure, assembly, and antigenicity of hepatitis B virus capsid proteins, *Adv. Virus. Res.*, **64**, 125–164.

134. Blanchard, E., Brand, D., Trassard, S., Goudeau, A., and Roingeard, P. (2002) Hepatitis C virus-like particle morphogenesis, *J. Virol.,* **76**, 4073–4079.

135. Chen, P. L., Wang, M., Ou, W. C., Lii, C. K., Chen, L. S., and Chang, D. (2001) Disulfide bonds stabilize JC virus capsid-like structure by protecting calcium ions from chelation, *FEBS Lett.*, **500**, 109–113.

136. Strable, E., Prasuhn, D. E., Udit, A. K., Brown, S., Link, A. J., Ngo, J. T., Lander, G., Quispe, J., Potter, C. S., Carragher, B., Tirrell, D. A., and Finn, M. G. (2008) Unnatural amino acid incorporation into virus-like particles, *Bioconjugate Chem.*, **19**, 866–875.

137. Chapman, M. S., and Liljas, L. (2003) Structural folds of viral proteins, *Adv. Protein Chem.*, **64**, 125–196.

138. Rossmann, M. G., and Johnson, J. E. (1989) Icosahedral RNA virus structure, *Annu. Rev. Biochem.*, **58**, 533–569.

139. Lucas, R. W., Larson, S. B., and McPherson, A. (2002) The crystallographic structure of brome mosaic virus, *J. Mol. Biol.*, **317**, 95–108.

140. Speir, J. A., Munshi, S., Wang, G., Baker, T. S., and Johnson, J. E. (1995) Structures of the native and swollen forms of cowpea chlorotic mottle virus determined by X-ray crystallography and cryo-electron microscopy, *Structure,* **3**, 63–78.

141. Douglas, T., and Young, M. (1998) Host-guest encapsulation of materials by assembled virus protein cages, *Nature,* **393**, 152–155.

142. Comellas-Aragonès, M., Engelkamp, H., Claessen, V. I., Sommerdijk, N. A., Rowan, A. E., Christianen, P. C., Maan, J. C., Verduin, B. J., Cornelissen, J. J., Nolte, R. J. (2007) A virus-based single-enzyme nanoreactor, *Nat. Nanotechnol.*, **2**, 635–639.

143. Minten, I. J., Hendriks, L. J., Nolte, R. J., and Cornelissen, J. J. (2009) Controlled encapsulation of multiple proteins in virus capsids, *J. Am. Chem. Soc.*, **131**, 17771–17773.

144. Krupovic, M., and Bamford, D. H. (2008) Virus evolution: how far does the double beta-barrel viral linage extend? *Nat. Rev. Microbiol.*, **6**, 941–948.

145. Butcher, S. J., Manole, V., and Karhu, N. J. (2012) Lipid-containing viruses: bacteriophage PRD1 assembly, in *Viral Molecular Machines*, eds. Rossmann, M. G., and Rao, V. B. (Springer, New York), 365–378.

146. Singh, R., and Kostarelos, K. (2009) Designer adenoviruses for nanomedicine and nanodiagnostics, *Trends Biotechnol.*, **27**, 220–229.

147. Wikoff, W. R., Liljas, L., Duda, R. L., Tsuruta, H., Hendrix, R. W., and Johnson, J. E. (2000) Topologically linked protein rings in the bacteriophage HK97 capsid, *Science*, **289**, 2129–2133.

148. O'Neil, A., Reichhardt, C., Johnson, B., Prevelige, P. E., and Douglas, T. (2011) Genetically programmed in vivo packaging ofprotein cargo and its controlled release from bacteriophage P22, *Angew. Chem., Int. Ed. Engl.*, **50**, 7425–7428.

149. Jiang, W., Li, Z., Zhang, Z., Baker, M. L., Prevelige, P. E., and Chiu W. (2003) Coat protein fold and maturation transition of bacteriophage P22 seen at subnanometer resolutions, *Nat. Struct. Biol.*, **10**, 131–135.

150. Huiskonen, J. T., de Haas, F., Bubeck, D., Bamford, D. H., Fuller, S. D., and Butcher, S. J. (2006) Structure of the bacteriophage phi6nucleocapsid suggests a mechanism for sequential RNA packaging, *Structure*, **14**, 1039–1048.

151. Aalto, A. P., Sarin, L. P., van Dijk, A. A., Saarma, M., Poranen, M. M., Arumäe, U., and Bamford, D. H. (2007) Large-scale production of dsRNA and siRNA pools for RNA interference utilizing bacteriophage phi6 RNA-dependent RNA polymerase, *RNA*, **13**, 422–429.

152. Carstens, E. B. (2010) Ratification vote on taxonomic proposals to the International Committee on Taxonomy of Viruses, *Arch. Virol.*, **155**, 133–146.

153. Stubbs, G., and Kendall, A. (2012) Helical viruses, in *Viral Molecular Machines*, eds. Rossmann, M. G., and Rao, V. B. (Springer, New York), 631–658.

154. Goulet, A., Blangy, S., Redder, P., Prangishvili, D., Felisberto-Rodrigues, C., Forterre, P., Campanacci, V., and Cambillau, C. (2009) Acidianus filamentous virus 1 coat proteins display a helical fold spanning the filamentous archaeal viruses lineage, *Proc. Natl. Acad. Sci. U. S. A.*, **106**, 21155–21160.

155. Smith, M. L., Lindbo, J. A., Dillard-Telm, S., Brosio, P. M., Lasnik, A. B., McCormick, A. A., Nguyen, L. V., and Palmer, K. E. (2006) Modified tobacco mosaic virus particles as scaffolds for display of protein antigens for vaccine applications, *Virology*, **348**, 475–488.

156. Werner, S., Marillonnet, S., Hause, G., Klimyuk, V., and Gleba, Y. (2006) Immunoabsorbent nanoparticles based on a tobamovirus displaying protein A, *Proc. Natl. Acad. Sci. U. S. A.*, **103**, 17678–17683.

157. Kendall, A., McDonald, M., Bian, W., Bowles, T., Baumgarten, S. C., Shi, J., Stewart, P. L., Bullitt, E., Gore, D., Irving, T. C., Havens, W. M., Ghabrial,

S. A., Wall, J. S., and Stubbs, G. (2008) Structure of flexible filamentous plant viruses, *J. Virol.*, **82**, 9546–9554.

158. Parker, L., Kendall, A., and Stubbs, G. (2002)Surface features of potato virus X from fiber diffraction, *Virology*, **300**, 291–295.
159. Carette, N., Engelkamp, H., Akpa, E., Pierre, S. J., Cameron, N. R., Christianen, P. C., Maan, J. C., Thies, J. C., Weberskirch, R., Rowan, A. E., Nolte, R. J., Michon, T., and Van Hest, J. C. (2007) A virus-based biocatalyst, *Nat. Nanotechnol.*, **2**,226–229.
160. Paschke, M. (2006) Phage display systems and their applications, *Appl. Microbiol. Biotechnol.*, **70**, 2–11.
161. Fischlechner, M., and Donath, E. (2007) Viruses as building blocks for materials and devices, *Angew. Chem., Int. Ed. Engl.*, **46**, 3184–3193.
162. Li, L., Jose, J., Xiang, Y., Kuhn, R. J., and Rossmann, M. G. (2010) Structural changes of envelope proteins during alphavirus fusion, *Nature*, **468**, 705–708.
163. Voss, J. E., Vaney, M. C., Duquerroy, S., Vonrhein, C., Girard-Blanc, C., Crublet, E., Thompson, A., Bricogne, G., and Rey, F. A. (2010) Glycoprotein organization of Chikungunya virus particles revealed by X-ray crystallography, *Nature*, **468**, 709–712.
164. Buchholz, C. J., Mühlebach, M. D., and Cichutek, K. (2009) Lentiviral vectors with measles virus glycoproteins: dream team for gene transfer, *Trends Biotech.*, **27**, 259–265.
165. Ernst, W. J., Spenger, A., Toellner, L., Katinger, H., and Grabherr, R. M. (2000) Expanding baculovirus surface display. Modification of the native coat protein gp64 of Autographacalifornica NPV, *Eur. J. Biochem.*, **267**, 4033
166. Ernst, W., Schinko, T., Spenger, A., Oker-Blom, C., and Grabherr, R. (2006) Improving baculovirus transduction of mammalian cells by surface display of a RGD-motif, *J. Biotechnol.*, **126**, 237.
167. Khare, P. D., Russell, S. J., and Federspiel, M. J. (2003) Avian leukosis virus is a versatile eukaryotic platform for polypeptide display, *Virology*, **315**, 303–312.
168. Schlehuber, L. D., and Rose, J. K. (2004) Prediction and identification of a permissive epitope insertion site in the vesicular stomatitis virus glycoprotein, *J. Virol.*, **78**, 5079–5087.
169. Kayman, S. C., Park, H., Saxon, M., and Pinter, A. (1999) The hypervariable domain of the murine leukemia virus surface protein tolerates large insertions and deletions, enabling development of a retroviral particle display system, *J. Virol.*, **73**,1802–1808.

170. Li, Z., Mueller, S. N., Ye, L., Bu, X., Yang, C., Ahmed, R., and Steinhauer, D. A. (2005) Chimeric influenza virus hemagglutinin proteins containing large domains of the Bacillus anthracis protective antigen: protein characterization, incorporation into infectious influenza viruses, and antigenicity, *J. Virol.*, **79**, 10003–10012.

171. Granieri, L., Baret, J. C., Griffiths, A.D and Merten, C. A. (2010) High-throughput screening of enzymes byretroviral display using droplet-based microfluidics, *Chem. Biol.*, **17**,229–235.

172. Roosnick, M. J. (2012) Plant virus metagenomics: biodiversity and ecology, *Annu. Rev. Genet.*, **46**, 357–367.

173. Prangishvili, D., Forterre, P., and Garrett, R. A. (2006) Viruses of the Archaea: a unifying view, *Nat. Rev. Microbiol.*, **4**, 837–848.

174. Prangishvili, D., Garrett, R. A., and Koonin, E. V. (2006) Evolutionary genomic of archaeal viruses: unique viral genomes in the third domain of life, *Virus Res.*, **117**, 52–67.

175. Ackermann, H. W., and Prangishvili, D. (2012) Prokaryote viruses studied by electron microscopy, *Arch. Virol.*, **157**, 1843–1849.

176. Atanasova, N. S., Roine, E., Oren, A., Bamford, D. H., and Oksanen, H. M. (2012) Global network of specific virus-host interactions in hypersaline environments, *Environ. Microbiol.*, **14**, 426–440.

177. Häring, M., Vestergaard, G., Rachel, R., Chen, L., Garrett, R. A., and Prangishvili, D. (2005) Virology: independent virus development outside a host, *Nature*, **436**, 1101–1102.

178. Mochizuki, T., Krupovic, M., Pehau-Arnaudet, G., Sako, Y., Forterre, P., and Prangishvili, D. (2012) Archaeal virus with exceptional virion architecture and the largest single-stranded DNA genome, *Proc. Natl. Acad. Sci. U. S. A.*, **109**, 13386–13391.

179. Pietilä, M. K., Laurinavičius, S., Sund, J., Roine, E., and Bamford, D. H. (2010) The single-stranded DNA genome of novel archaeal virus *Halorubrum* pleomorphic virus 1 is enclosed in the envelop decorated with glycoprotein spikes, *J. Virol.*, **84**, 788–798.

Chapter 2

Strategies for Virus–Enzyme Coupling

Daniela Cardinale, Noëlle Carette, and Thierry Michon
Equipe de Virologie, UMR 1332 Biologie du Fruit et Pathologie, INRA et Université de Bordeaux, CS 20032, 33882 Villenave d'Ornon Cedex, France
michon@bordeaux.inra.fr

2.1 Covalent Attachment of Enzymes to Virus Nanoparticles

It is out of the scope of this section to give an exhaustive presentation of all possible methods aiming at assembling enzymes (and more generally proteins) with virus nanoparticles (VNPs). Processes allowing a controlled positioning of the enzyme on or into a VNP, while preserving its catalytic activity, require bioconjugation strategies of the highest specificity available. Within the last 15 years, the combination of progresses in bio-organic chemistry and genetic engineering has led to very sophisticated bioinspired achievements, which pave the path for future developments. We will focus on these strategies and explore their potentials and limitations. A general rationale for choosing between various coupling strategies was recently proposed [1]. In addition to this section, the reader might find useful to consult it.

Enzyme Nanocarriers
Edited by Daniela Cardinale and Thierry Michon
Copyright © 2016 Pan Stanford Publishing Pte. Ltd.
ISBN 978-981-4613-42-2 (Hardcover), 978-981-4613-43-9 (eBook)
www.panstanford.com

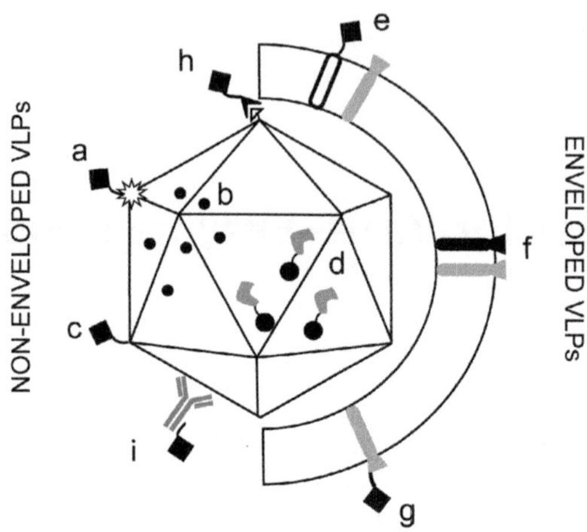

Figure 2.1 Outline of ENC decoration strategies described in the chapter. (a) Chemical modification, (b) confinement (encapsidation), (c) direct fusion with CP (nonenveloped particles and core-like particles), (d) guided encapsidation, (e) membrane anchoring, (f) pseudotyping, (g) fusion with envelope proteins, (h) interacting domains, and (i) immunostrategies.

By way of introduction, here are a number of general considerations to keep in mind when defining appropriate strategies.

- *Prevention of homopolymers' formation.* When classical reactions between amino acid side chains are involved, both parts (i.e., VNPs and enzymes) can display the same kind of reactive moiety, giving rise to a population of heteropolymers (VNP-Ez) contaminated with undesired homopolymers species (VNP–VNP, Ez–Ez). A careful tuning of the ratio between the reactive species may orientate favorably the species produced. However, in solution nano-objects of several tenths of nanometers in diameters bearing many possible reacting moieties on their surface are supposed to behave very differently from proteins whose hydrodynamic radius rarely exceeds 10 nm. The use of a different type of reactive moiety on each partner should prevent the appearance of homopolymers. Heterobifunctional coupling reagents bearing a combination of amine-reactive and photoactivable phenyl azides groups are the most suitable for these applications. In these two-step reactions, one of the partners is first functionalized, separated from by-products and unmodified forms, and mixed in the appropriate proportions with the other partner for the coupling step (Fig. 2.2). Other crosslinkers of this type reacting with –COOH and –SH groups carried by amino acid side chains are commercially available. It appears that many of them, although quite selective, often give poor yields, impairing the isolation of a significant amount of functionalized species. These reagents have been mainly used for analytical purposes such as interaction site mapping or biological complexes [2, 3]. But crosslinkers required for coupling two molecules displaying no natural affinity need to be longer than those used for subunit crosslinking within biological native complexes. The spacer length introduced by the coupling agent should be considered case to case [4]. Finally, the use of these coupling reagents for production of functional biomaterials is beyond a reasonable cost.

- *Preservation of enzyme activity after coupling reactions.* The choice of a chemistry allowing coupling in mild biocompatible conditions and a careful design of coupling sites well targeted on the virus and enzyme surface will ensure both a positional control of the enzyme on the scaffold and a preservation of the activity.
- *Chemoselectivity.* Enzymes can be potentially coupled to many sites on the VNP. Tunning the balance between chemoselectivity and full conversion remains problematic. In an attempt to reach full conversion, the addition of excess reagent often leads to overmodification and in turn to a loss of chemoselectivity. On the other hand, maintaining selectivity often leads to low yields and poor separation of the modified protein from contaminations. It is always necessary to perform a systematic study to determine the enzyme suitable to viral particles (VPs) for a good coupling yield (as sometimes more is not better) leading to a satisfying end product.
- *Site-specific coupling.* As the aim considered in targeting of grafting sites on the protein surface is to functionalize VNPs with enzymes, choosing the site specificity of the bioconjugation according to criteria such as protein orientation or catalytic and binding sites positioning on the virus surface should, in principle, preserve the catalytic activity. In the ideal situation, a precise positioning of enzymes on VNPs requires the

identification of accessible coupling sites both on the particle and on the enzyme. In addition the coupling site on the enzyme must be chosen so that the substrate binding remains unaffected. Knowledge of the 3D structures of enzyme and VNP surfaces helps find out potential good sites. Site-directed mutagenesis is the only way to install a single copy of a new functional group in a single location on the protein surface. Depending on the nature of the virus, several strategies can be derived for VNP modification. When bacterial viruses (i.e., bacteriophages) are used, amino acid substitutions can be introduced by site-directed mutagenesis into the virus genome. Consequently, VPs harboring the suitable grafting site are expected to be naturally produced in vivo and subsequently purified prior to modification. A successful strategy of this kind involving bacteriophage Qβ is detailed below [5]. In the case of eukaryotic viruses, VPs can sometimes be reassembled in vitro from coat protein units overexpressed in heterologous systems (i.e., bacteria). This is the case of cowpea chlorotic mottle virus (CCMV)-like particles, which were used as enzyme nanocarriers (ENCs) [6–9]. Some examples are presented in Chapter 3 of this book by experts in the field. If the VP of interest cannot be reassembled from isolated coat protein units, particles must be prepared from extracts of the infected host organism. In such a case, if a suitable amino acid needs to be introduced for grafting, the required site-directed mutagenesis has to be performed into the virus genome. The engineered virus is then inoculated to the host and must have retained its infectious power. The purified VNPs are subsequently crosslinked with the enzyme in vitro. Such a strategy was used to introduce a lysine residue on the surface of tobacco mosaic virus (TMV). Knowing the precise molecular structure of the VP, a site supposed to satisfy all topological requirements was identified. However, this recombinant virus was suboptimal; the yield of purified VPs was low (0.1–0.2 mg/g infected tissue) and aggregated upon storage. Subsequently, a lysine was introduced at the *N*-terminus of the TMV coat protein, with three randomized codons placed immediately upstream. The library obtained was screened in planta for the selection of stable TMV particles [10].

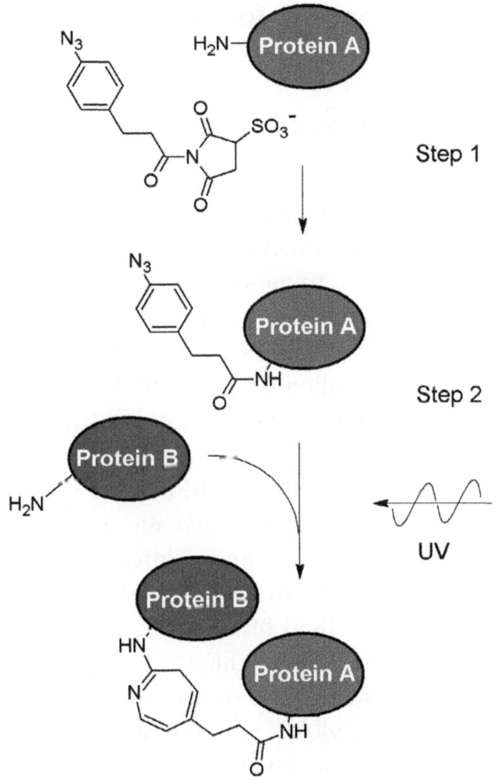

Figure 2.2 General scheme for protein coupling by bifunctional reagents. Bifunctional reagents bearing a sulfo-*N*-hydroxysuccinimidyl (NHS) ester group and a photoreactive phenyl azide group interspaced by arms of various lengths and functionalities are commercially available. In step 1, the NHS ester specifically reacts with the available primary amines of the first protein. In step 2, (after purification) the modified protein is incubated with the second protein. The sample is exposed to UV light to photocrosslink the two proteins. It is to be mentioned that the phenyl azide moiety reacts unspecifically with the second protein. Upon exposition to UV, a nitrene group is formed that can initiate addition reactions with double bonds, insertion into C–H and N–H sites, or ring expansion, which reacts with a nucleophile (e.g., primary amines). When primary amines are available, the latter reaction dominates. As thiol-containing reducing agents (e.g., DTT or 2-mercaptoethanol) reduce the azide functional group to an amine, they must be avoided in the sample solution. Reactions must be performed in amine-free buffer conditions compatible with the first step reaction.

2.1.1 Protein Engineering for Targeting Specific Coupling Sites

2.1.1.1 Natural amino acids

A simple and efficient way to covalently couple proteins makes use of the natural reactivity of the amino acid side chains exposed at their surface. Methods have been successfully developed for bioconjugation of VNPs with many compounds. The reader might find it useful to consult Refs. [11, 12] for recent overviews. In attempts to obtain a positional control of enzymes on VNPs, the amino acid reactivity needs to be strictly controlled. Many examples show that a directed bioconjugation through amino acids pre-existing on the protein partners is difficult. Although primary amines display high reactivity toward *N*-hydroxysuccinimide (NHS) esters or isothiocyanates, the large number of lysine residues usually present in the protein prevents a positional control of the modification. Similarly, the use of carbodiimides and amines to capture carboxylates is precluded. More chemoselective are thiol functions. Cysteine residues can be introduced at specific positions within the protein and subsequently alkylated with various maleimides or bromo/iodo acetamides derivatives. The peculiar reactivity of tyrosine can lead to useful alternatives and deserves to be discussed.

2.1.1.1.1 *Cysteine*

In the ideal situation, selective coupling may be induced by introducing a specific functional group at a chosen position on each of the two partners (capsid protein [CP] and enzyme). If, for instance, each of the two proteins can be efficiently expressed in *Escherichia coli*, site-directed mutagenesis can be applied to introduce infrequent amino acids at suitable sites. Owing to the low abundance of cysteine in most protein sequences, disulfide bridges provide the most common way to achieve site-selective protein coupling. However, assigning a cysteine carrying a thiol function on each side provides a homobifonctional system, which suffers from several drawbacks. First, disulfide bridge formation is governed by oxidative conditions often difficult to control, especially if the protein system of interest is only stable in an alkaline range of pH 8–9, where disulfide formation is favored. Second, coupling will give rise to both heterodimerization

(VNP–enzyme) and homodimerization. A strategy to avoid this later requires activation of the thiol, preventing disulfide bridge formation. *o*-Mesitylenesulfonylhydroxylamine (MSH) is a potential reagent in the oxidative elimination of Cys to dehydroalanine (Dha). To achieve bioconjugation, a disulfide bridge between the Dha moiety on the enzyme and a free cysteine on the VP surface can be subsequently formed (Fig. 2.3). Although oxidative elimination of cysteine is expected to be faster than methionine since the former

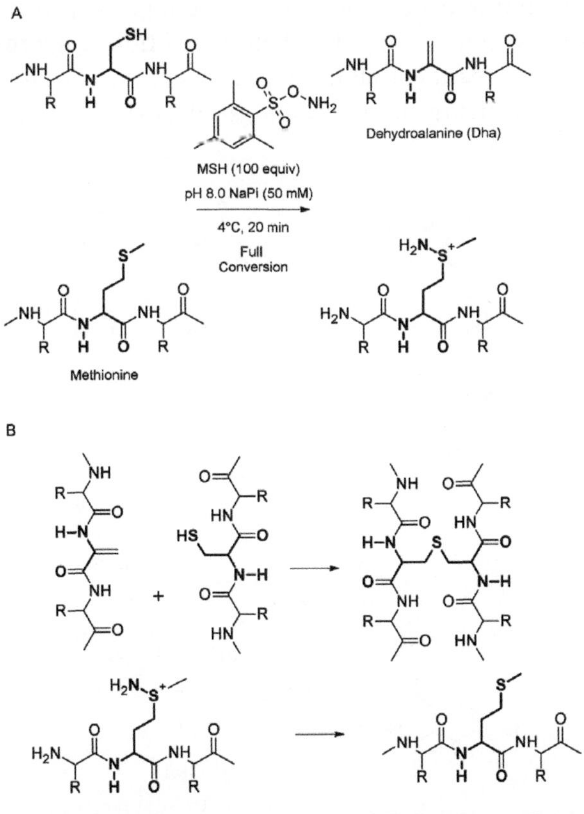

Figure 2.3 (A) Conversion of cysteine to dehydroalanine by *o*-mesitylenesulfonylhydroxylamine (MSH). Although methionine residues also react with MSH, the by-product is unstable and fast reverts to methionine. (B) Dehydroalanine efficiently condenses with cysteine, giving a disulfide bound. The chemical bounds belonging to target amino acids are in bold.

proceeds through the removal of an acidic α proton, while the latter through the less labile β proton, methionine reacts with MSH also. When the Dha derivative and the partner bringing the free thiol are mixed, a disulfide crosslinking is favored and methionine residues are regenerated. The end product analysis shows that the method does not induce significant modifications of methionine residues [13].

In case there is no available cysteine accessible on the VP surface, *N*-succinimidyl S-acetylthioacetate (SATA), a thiolation reagent, can be used to introduce a protected thiol group on primary amino groups (lysine side chains). Just before addition, a hydroxylamine hydrochloride treatment allows to deprotect the introduced thiol, which is subsequently used for conjugation [14] (Fig. 2.4).

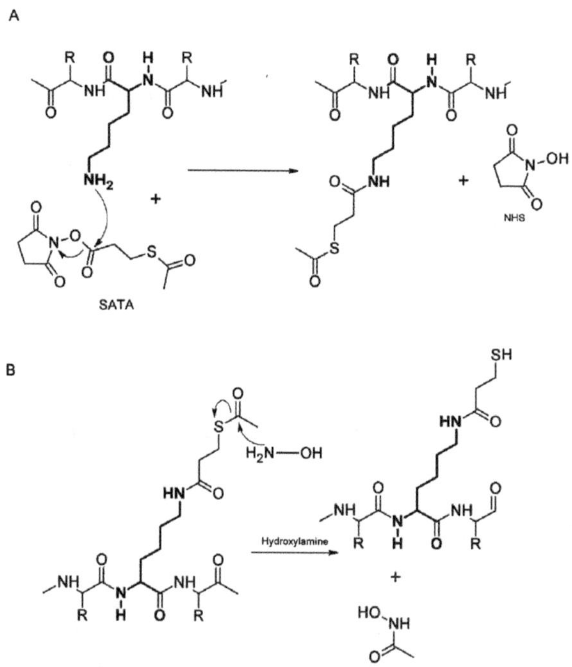

Figure 2.4 Conversion of lysine residues to cysteine using *N*-succinimidyl S-acetylthioacetate (SATA). (A) The nucleophilic attack of SATA carbonyl by primary amines results in the transfer of a thioester on the lysine side chain through an amide bound with the release of the *N*-hydroxysuccinimide (NHS) moiety. (B) The thioester is hydrolyzed by alkaline treatment prior to reaction with the second partner.

2.1.1.1.2 *Lysine*

Amine targeting is the quickest and most reliable way to produce a protein bioconjugate. But the very high prevalence of lysine groups on virtually all protein surfaces leads to the inability to control the number of modifications and the specific locations in which they occur. However, the favorable exposure of a given amino acid on the surface of the VNP can be exploited with profit for the coupling of enzymes. Indeed many amino acids accessible at the surface of monomeric CPs are buried upon the supramolecular assembly of the monomers into the capsidic edifice. For instance, only four lysine residues out of seven are exposed to the bacteriophage Qβ surface and all are very reactive [15]. The cowpea mosaic virus (CPMV) capsid is an icosahedral particle (30 nm diameter). It is formed by 60 copies of an asymmetric unit comprising a small (S) and a large (L) subunit. CPMV cannot easily be obtained from heterologous expression in *E. coli*, but gram quantities can be isolated from a kilogram of infected leaf tissue. The infectious cDNA clones of the virus are made under the control of the 35S promoter for efficient and convenient transfection into plants [16], which enables the surface properties of the CPs to be altered through site-directed mutagenesis. The asymmetric unit contains five lysine residues. A specific coupling was targeted by a systematic replacement of all lysine residues but one. All of the mutants were viable, and VPs were recovered from infected plants. The replacement allows a fine evaluation of the reactivity of each of the five amino groups on the VP surface, demonstrating strong environmental effects [17]. The specific coupling of various molecular species, including proteins, on the surface of CPMV through lysine targeting was reported. A mutant of T4 lysozyme carrying a single cysteine residue was coupled to the CPMV surface through succinimidyl 4-(*N*-maleimidomethyl) cyclohexane-1-carboxylate (SMCC), a heterobifunctional crosslinker, allowing the coupling of amines to thiol groups (Fig. 2.5). NHS esters are prompt to hydrolyze in solution, and it proved to be more efficient to proceed to the lysine modification of the virus first, followed by the maleimide reaction on the lysozyme cysteine. Virtually all the targeted amines were coupled to lysozyme, and the chemical crosslinking procedure did not affect the biological activity of the conjugated enzyme adversely [18].

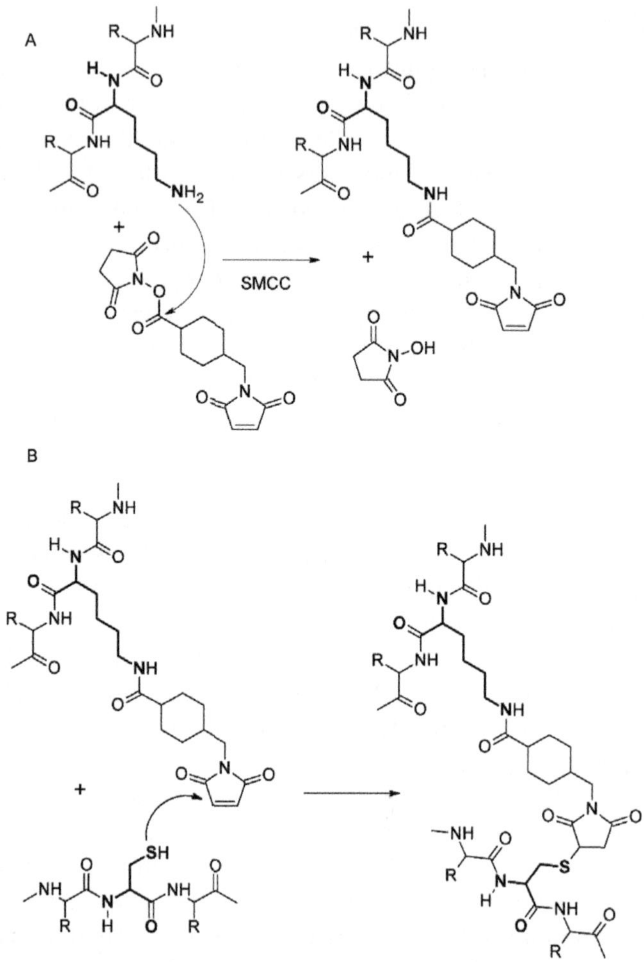

Figure 2.5 Lysine–cysteine crosslinking. Use of the heterobifunctional crosslinker succinimidyl 4-(*N*-maleimidomethyl)cyclohexane-1-carboxylate) (SMCC). (A) Fast reaction of the NHS ester with lysine primary amines. (B) Addition of the –SH group to the maleimide moiety.

With a view to generate coupling targets at the VP surface, parts of the CPs that are exposed can be genetically extended with peptides containing the reactive amino acid. However, the addition of empirically designed short peptides may affect surface properties, compromising the VP assembly. To overcome these problems, a screening scheme to select for viruses displaying the residue of interest, while retaining the property of self-assembly and infectivity, can be applied [10].

For instance, a lysine residue was introduced at the *N*-terminus of the TMV CP, with three randomized codons placed immediately upstream. Transcripts from approximately 3600 clones were pooled and inoculated onto *Nicotiana benthamiana* plants to generate a library of TMV CP mutants. After the selection, 11 unique CP fusions bearing the lysine residue, while keeping self-assembly properties, were identified. All had at least one acidic residue among the three random residues preceding the fixed lysine residue. Together, this small collection of selected viruses suggests that in this particular case, one or more surface-exposed acidic residues are needed to mitigate the presence of the surface-exposed lysine. For an extensive discussion of lysine reactivity at the surface of VPs see Ref. [11].

2.1.1.1.3 *Tyrosine*

Tyrosine residues, although more frequent than their cysteine counterparts, are also poorly represented in proteins. Their unique reactivity could make them attractive species for coupling VPs to enzymes. The complex of NiII with the tripeptide NH_2–glycine–glycine–histidine–COOH (NH_2-GGH) can mediate the crosslinking of proximal proteins through the formation of a tyrosine–tyrosine bond by an oxidatively triggered reaction [19, 20]. In this way, adjacent coat proteins within the CPMV were crosslinked through dityrosine formation [21]. However, this only occurs between two protein partners, which naturally pre-exist in a binary complex, and this strategy cannot be applied as a general scheme to couple VPs and enzymes. Peroxidases efficiently catalyze the extraction of a single electron from the aromatic ring of tyrosine residues in water. Tyrosine radicals will in turn condense to form dityrosine [22]. This enzymatically assisted crosslinking is quite appealing as it is performed in very mild conditions, allowing the preservation of biological structures. To form a complex between noninteracting proteins, it is expected that tyrosine–tyrosine crosslinking will only occur if tyrosine residues are accessible from the surface. This is not the most frequent situation. Because of the weakly polar feature of the phenolic ring, the tyrosine side chain tends to be buried in the hydrophobic core of the protein. In such a case, the binding of the tyrosinyl substrate to the peroxidase occurs at low frequencies. We used a fungus peroxidase, manganese-dependent peroxidase from *Phanerochaete chrysosporium* (MnP), to induce crosslinking between poorly accessible tyrosine residues. MnP does not need to bind tyrosine directly to produce its oxidation. This enzyme acts according to "indirect" catalysis. Starting from H_2O_2, lactate, and Mn^{2+}, MnP produces a [Mn^{3+}–lactate] complex that diffuses freely in the reaction mixture and is able to oxidize poorly exposed tyrosine residues (Fig. 2.6A). The yield of dityrosine was increased fivefold with respect to what is obtained using horseradish peroxidise, which needs to complex with the phenolic ring to convert it into its radical form [23]. One should also consider that the phenolic

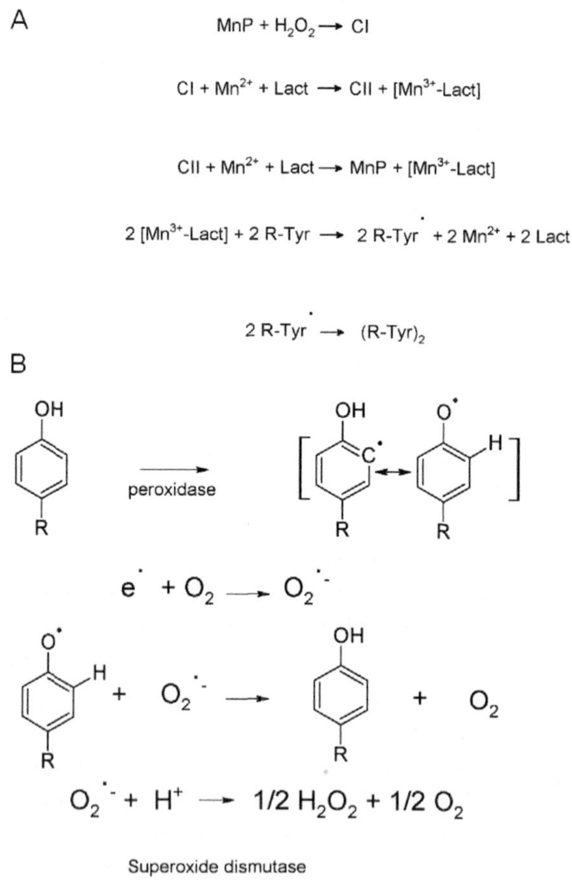

Figure 2.6 (A) Mechanism of MnP-catalyzed oxidation of tyrosine. CI, MnP complex 1; CII, MnP complex II. (B) Neutralization of superoxide ions by superoxide dismutase.

radicals formed are highly reactive and their lifetime is short. Most of them are reduced by superoxide ions before they condense. We observed that by adding superoxide dismutase, which consumes superoxides, the lifetime of tyrosine radicals is increased, improving the dityrosine yield (Fig. 2.6B). Two phenol radicals require to be close enough (less than 4 Å) to be coupled [21]. Surface-accessible tyrosines can be added to VPs and enzymes. The Bolton–Hunter reagent, succinimidyl-3-(4-hydroxy-phenyl) propionate (SHPP), appends tyrosine-like side chains to surface-exposed lysine residues. The phenolic rings added that way are targeted significantly better [24] (Fig. 2.7).

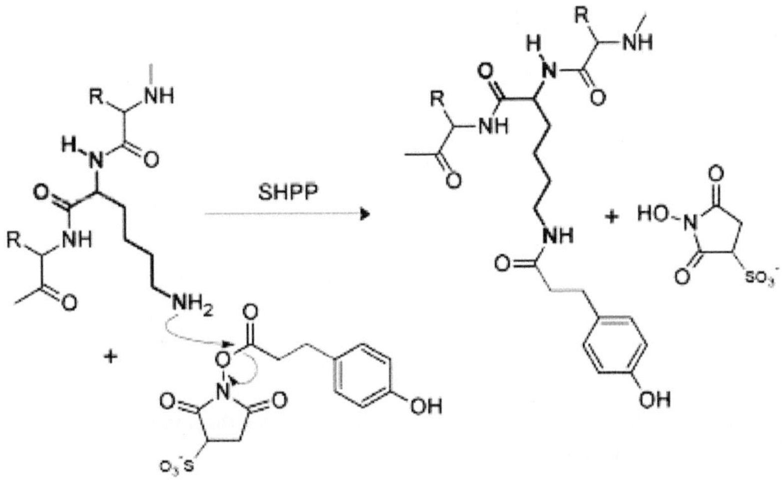

Figure 2.7 Transfer of a tyrosine group to the end of a lysine side chain by the nucleophilic attack of sulfo-SHP.

2.1.1.1.4 *Aspartic and glutamic acids*

An activation step with 1-ethyl-3-(3-dimethylaminopropyl) carbodiimide (EDC) is generally used, which allows the nucleophilic attack of the carbonyl by primary amines (Fig. 2.8). It is very difficult to direct the reaction toward the formation of heterodimers (i.e., VP–enzyme), carboxylates tending to react with primary amines within the same molecular species. This leads in particular to the aggregation of VPs.

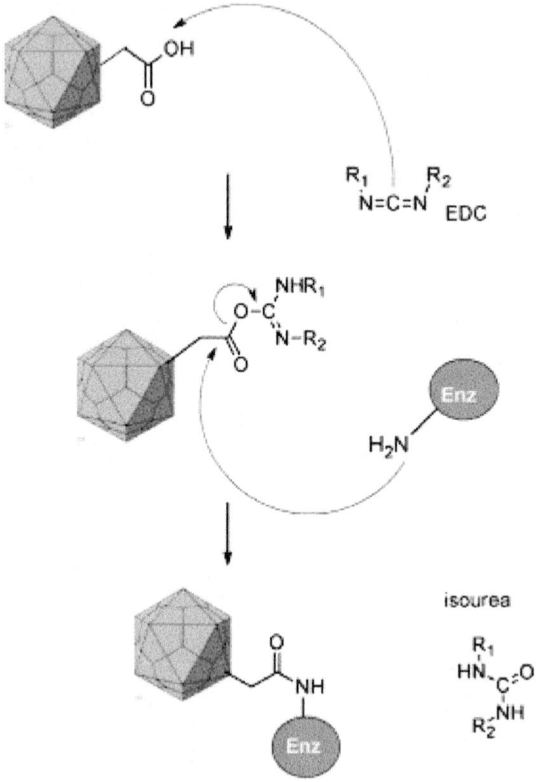

Figure 2.8 Carboxylate activation by EDC and its subsequent nucleophilic attack by a primary amine.

In their book *Viral Nanoparticles, Tools for Materials Science and Biomedecine*, Steinmetz and Manchester give an overview of the various functionalities that have been attached to VPs [12]. Interestingly, at most 5% of the whole functionalization of VPs involves enzymes. Because the activity of the enzyme needs to be preserved, we focused our survey on the mildest possible chemistry, while trying to preserve selectivity. The various strategies exposed have been summed up in Fig. 2.9.

All coupling strategies involving the chemistry presented above use the natural reactivity of amino acid residues. Although operated in mild conditions compatible with protein stability they require working with an excess of precious functionalized proteins with respect to VPs and time-consuming subsequent purification steps.

Reports employing thiol–maleimide chemistry for the attachment of proteins (up to 22 kDa) to CPMV required the use of a 50-fold excess of protein with respect to the viral asymmetric unit, as well as subsequent chromatographic purification of the desired conjugate [18].

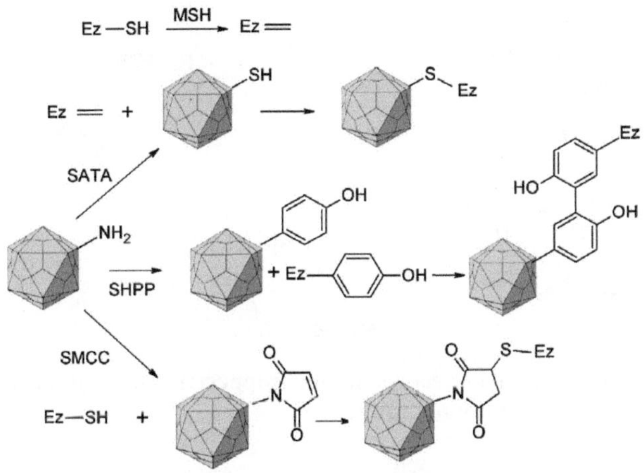

Figure 2.9 Strategies to improve the selectivity of the bioconjugation via the reactivity of natural amino acids.

As shown below, new routes of biocompatible crosslinking are now being explored using the reactivity of non-natural chemical moieties incorporated into proteins. This expends the tool box available for protein chemistry. These approaches are friendlier and allow a better positional control of the coupling.

2.1.1.2 Copper(I)-catalyzed azide–alkyne cycloaddition

The reaction of terminal or internal alkynes with organic azides has been known for more than a century. It proceeds via cycloaddition, producing a heterocycle involving 1,2,3 triazole. However, its high activation barrier results in exceedingly low reaction rates for unactivated reactants, even at elevated temperature. The use of copper as a catalyst increased the rate of the reaction by a factor of 107 relative to the thermal process, making it conveniently fast at room temperature. For a detailed analysis of copper(I)-catalyzed azide–alkyne cycloaddition (CuAAC) the reader is referred to Ref. [25]. A reaction scheme is presented in Fig. 2.10.

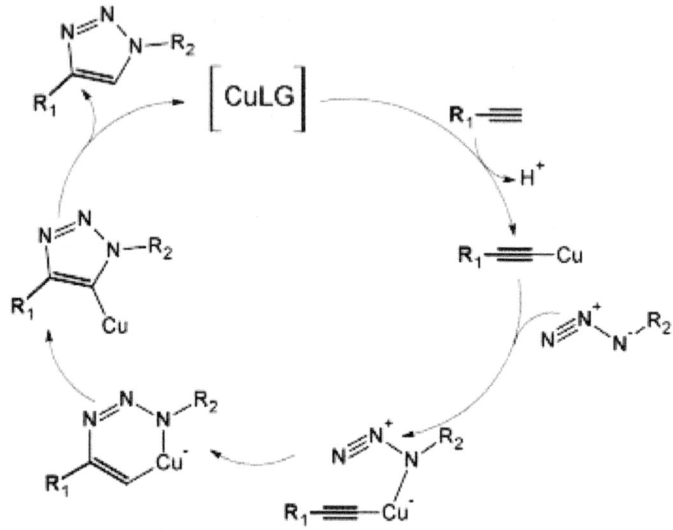

Figure 2.10 Proposed mechanism for copper(I)-catalyzed azide–alkyne cycloaddition (CuAAC).

The reaction proceeds in many protic and aprotic solvents, including water, and is unaffected by most organic and inorganic functional groups, eliminating the need for protecting group chemistry. CuAAC can be considered a regioselective "ligation" of azides and terminal alkynes. These reactions in mild conditions are characterized by high yields [26, 27]. Typically, the reactions require simple or no workup, or purification, of the product. A copper-binding ligand is used to accelerate the reaction, minimizes the oxidation of copper from +1 to +2 states, and prevents the metal from inducing protein aggregation or degradation [28]. Although it makes the catalytic system sensitive to oxygen and requires to be handled in an anaerobic environment, sulfonated bathophenanthroline (Fig. 2.11) is fully water soluble and makes a faster catalyst [29].

Prior to cycloaddition, azide and alkyne groups can be grafted on to capsids and enzymes by means of their natural reactivity (amine, thiol) following the strategies presented above. For instance, CPMV capsids were labeled with azides at surface-exposed lysine residues using standard NHS ester chemistry (Fig. 2.5). High concentrations of human transferrin (20 mg/mL), an 80 kDa iron transporter, were incubated with 15 eq of a maleimide–alkyne linker to afford the alkyne-derivatized protein with attachments made at accessible cysteines (and, to a lesser extent, lysine residues). The CPMV–transferrin conjugate was prepared by reaction of a CPMV azide functionalized with alkyne-derivatized transferrin using the [Cu(I)-sulfonated bathophenanthroline] complex as a catalyst. Simple centrifugation over a sucrose gradient allowed the recovery of modified VNPs. A significant number of transferrin molecules were arrayed on the particles. The virus–protein conjugates were isolated as individual particles, with no evidence of aggregation that might be expected if transferrin species bearing more than one alkyne were to couple to polyvalent CPMV azides. Transferrin molecules were covalently attached evenly over the CPMV surface. This procedure requires only a sixfold excess of transferrin and an easy recovery step with yields far superior to those possible with thiol–maleimide chemistry [29].

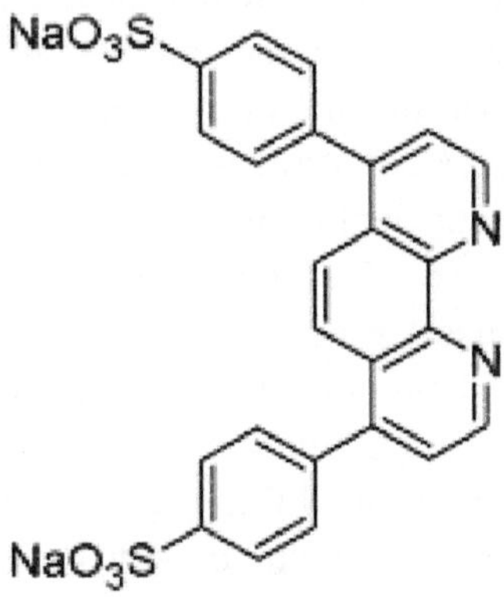

Figure 2.11 Sulfonated bathophenanthroline.

2.1.1.3 Unnatural Amino Acids

2.1.1.3.1 *Introduction*

In living organisms the translation machinery is permissive toward the incorporation of artificial amino acids into proteins. Structural analogues bearing nonbiological fluorinated [30] or unsaturated [31] side chain functions have been incorporated in vivo into proteins, in place of their natural counterparts. This opens up possibilities for an enlargement of proteins' reactivity, including new types of residues crosslinking. The most important characteristic of the CuAAC reaction is its unique bio-orthogonality, as neither azide nor terminal alkyne functional groups are generally present in natural systems. Consequently, the in vivo incorporation of azides and terminal alkynes into proteins can lead to very specific coupling. The success of this method relies on the recognition of analogues by the aminoacyl–tRNA synthetases, which safeguard the fidelity of amino acid incorporation into proteins. Methionyl–tRNA synthetase proved to be tolerant toward incorporation of methionine structural analogues such as ethionine, selenomethionine, telluromethionine, trifluoromethionine, and norleucine. It was later demonstrated that azidohomoalanine [32] and homopropargylglycine can also be incorporated with a good yield [33].

2.1.1.3.2 *Methods*

We will briefly describe the principle of the procedure. To avoid competition between methionine and its analogues, a bacterial strain unable to synthesize methionine (auxotrophy) is used. This methionine auxotroph *E. coli* strain is transformed with an expression vector bearing the gene coding for the protein of interest (enzyme or CP) under the control of an inducible promotor. In addition, the expression plasmid also contains the gene encoding methionyl–tRNA synthetase as it was shown that an efficient analogue incorporation requires to increase the amount of corresponding amino acyl–tRNA synthetase in the bacterium [33]. Cultures of bacteria in an M9 (minimum) medium supplemented with the 20 amino acids are grown to an exponential phase. The cells are then washed three times with M9 (without amino acids added), suspended in an M9 medium containing 19 amino acids plus the analogue, and the expression of the protein of interest is induced. The cells are harvested and treated for protein isolation. Although the substitution is never complete owing to an endogenous turnover of the pool of free methionine provided by proteins hydrolysis, a high yield of incorporation is obtained (above 70% of the methionine is usually replaced by the analogue). As the approach is not site specific, it leads to an extensive replacement of canonical amino acids with structural analogues. The incorporation of unnatural amino acids (UAAs) at multiple sites within proteins may impair the VNP assembly and/or induce misfolding and inactivation of the enzymes of interest. Mutagenesis is required to replace methionine at positions were modifications are unwanted. This is time consuming, and high incorporation can compromise the 3D structure of the protein. This approach requires a case-to-case examination. The CPs of bacteriophage Qβ spontaneously self-assemble into noninfectious pseudoparticles or virus-like particles (VLPs) in bacteria. In cases where methionine in the CP is replaced by azidohomoalanine or homopropargyl glycine, the CPs retain the ability to assemble into VLPs. These analogues are chemically inert under physiological conditions. The VLPs functionalized with these artificial amino acids can be isolated and coupled with the desired protein in vitro by CuAAC. Using the technology described above, a positional control of transferrin molecules on the VLP surface was targeted by applying a suitable site-directed mutagenesis in the gene encoding the CP to remove or insert methionine. The CP of bacteriophage Qβ possesses only the *N*-terminal methionine. An additional methionine was introduced into the protein either in place of the external lysine 16 (K16M) or in place of threonine 93 (T93M), which is exposed within the cavity; between 70% and 95% of the methionine residues were replaced. On Qβ VLPs (K16M) half of azidohomoalanine at position 16 were coupled with *N*-propargyl-transferrin. This corresponds to about 90 molecules of transferrin per particle. The same approach was applied to hepatitis B virus (HBV). The analogue incorporation into CPs and their self-assembly into VLPs was similar to what was observed with Qβ VLPs. However, attempts to perform CuAAC of various alkyne groups on the surface of the HBV VLPs induced particle dissociation. The nature of the ligand made no difference to particle stability. VLPs were recovered in yields inversely proportional to the amount of the CuAAC reaction that had occurred. The triazole linkage, which is common to each of the grafted groups, might be responsible for the instability of the virus-like HBV particle [5]. These comparative results show that many factors affect the stability of the functionalized particles. Bacteriophages and some plant viruses (e.g., CCMV) are preferred as icosahedral VPs because their CP can be engineered in the bacterium and self-assembled into VLPs in the absence of the virus genetic material. In summary, the two-step process of (i) genetically directed incorporation of azides and alkynes and (ii) CuAAC ligation is a generic method to couple virtually any type of enzyme to VLPs.

2.1.1.3.3 *Site-specific incorporation of unnatural amino acids into proteins*

Efficient systems have been developed to target the site-specific incorporation of UAAs, enabling a perfect positional control of the coupling (Fig. 2.12). An engineered tRNATyr/TyrRS pair derived from *Methanococcus jannaschii* has been used to site specifically incorporate UAAs into proteins in *E. coli* in response to the amber codon UAG (nonsense codon in *E. coli*) [34].

The illustration was made above that a smart combination of genetic engineering and of CuAAC in water is paving the way for future highly sophisticated achievements. Most enzymes are fragile, and the best technologies for their engineering involve genetic strategies. In the next sections some of these strategies such as coat protein–enzyme genetic fusions, reversible biomimetic interactions between enzymes and the VPs' surface, and the reprogramming of bacteria for the in vivo preparation of functionalized VPs are presented.

2.2 Genetic Strategies for Direct Enzyme Fusion

In this section various genetic strategies are presented that have been employed to successfully decorate VNPs with enzymes/proteins.

As described in Chapter 1, nature offers an abundant library of viruses with various structures and properties. Although they are not all suitable as protein nanocarriers, there is still a comfortable choice, depending on what is planned. For instance, empty shells (VLPs) issuing from plant icosahedral virus (spherical-like) can be assembled starting from functionalized monomeric CPs produced in bacteria. They may be used for functional material engineering. Enveloped viruses from animals will serve as drug vectors for cell therapy. Various examples will be first presented illustrating the use of nonenveloped viruses. A part will follow illustrating technologies using enveloped viruses.

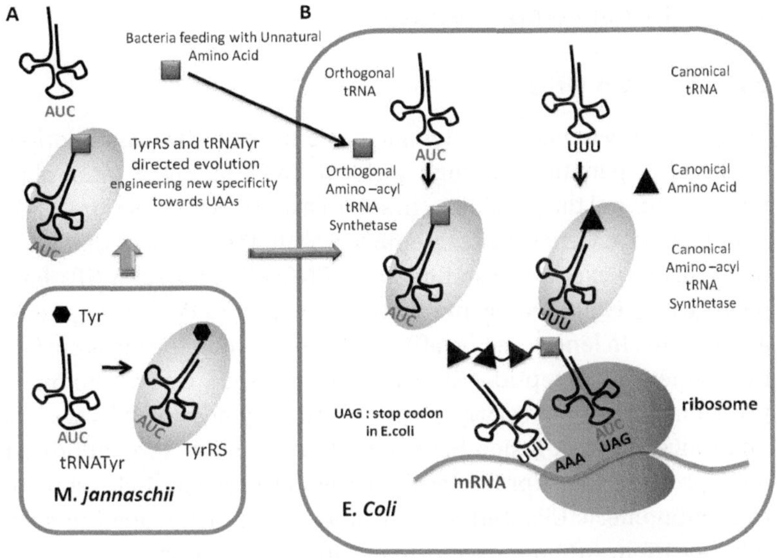

Figure 2.12 Strategy for site-specific incorporation of unnatural amino acids into proteins in *E. coli*. Expanding the genetic code of *E. coli* requires a suppressor tRNA that is not acylated by *E. coli* synthetases (orthogonal suppressor tRNA) and a synthetase that does not acylate any *E. coli* tRNA but efficiently charges the orthogonal suppressor tRNA. (A) UAG is a stop codon in *E. coli*, and protein synthesis stops whenever this codon is found on the mRNA sequence. *Methanococcus jannaschii* bears a tRNATyrCUA/TyrRS couple orthogonal to *E. coli*. Hence a tyrosine is linked to the ongrowing polypetide chain each time a UAG codon is red on the mRNA. A directed evolution procedure was applied to *M. jannaschii* tRNATyrCUA and TyrRS-mutated populations. On the one hand, a mutRNATyrCUA, a poor substrate for *E. coli* endogenous synthetases but still efficiently aminoacylated by the *M. jannaschii* TyrRS, was selected. On the other hand, an inactive TryRS for Tyr loading but providing an efficient acylation of mutRNATyrCUA by *o*-methyl-L-tyrosine was obtained. (B) After the introduction in *E. coli* of the genes coding for the mutRNATyrCUA/mutTyrRS couple, polypeptide chains were grown with site-specific incorporations of *o*-methyl-L-tyrosine each time the translation machinery was reading a UAG codon on the mRNA.

2.2.1 Nonenveloped Viruses

2.2.1.1 Plant viruses

The possibility to fuse peptides and proteins to CPs by manipulations of the viral genome, the high yield of VNPs obtained after plant inoculation, and the possibility in some cases (it is the case for CPMV) to induce an in vitro capsid assembly are suitable features that make plant viruses excellent candidates as ENCs. Plant virus VNPs have been largely exploited for production of vaccines. Peptides up to 50 amino acids in length can be efficiently exposed on the virus surface by engineering a peptide–CP genetic fusion [35]. One of the major problems in displaying large proteins and enzymes on VNPs is steric hindrance. A direct fusion between CPs and a bulky foreign protein often prevents the supramolecular assembly of the particles starting from monomeric CPs. Different strategies were developed to solve this question; attempts were made involving inserting a peptide spacer between the CP and the protein to be displayed. Infective nanoparticles displaying a 133-amino-acid-long functional fragment of protein A were obtained by inserting two different types of linkers, a flexible glycine-rich linker $(GGGGS)_3$ or a helical linker peptide $(EAAAK)_3$, between the CP of turnip vein clearing virus (TVCV) and the fusion protein. The addition of these linkers enabled correct folding of the fusion protein and CP domains, allowing the assembly of functionalized VNPs in planta. The packing of protein A was highly dense, with more than 2100 copies per particle, almost reaching the number of CPs forming the viral shell [36]. The $(GGGGS)_3$ linker was used to decorate potato virus Y (PVY) with rubredoxin, a small protein of 54 amino acids that was fused at the *N*-terminus of the CP. Fusion proteins were expressed in *E. coli*, giving rise to assembled VNPs [37]. The high density of proteins obtained in the above examples was unexpected. TVCV and PVY are high-aspect-ratio filamentous viruses, several hundreds of nanometers in length and about 15 nm in diameter. The tight packing of the CPs makes them less prone to expose foreign proteins on neighboring subunits than spherical VNPs. An interesting approach to facilitate protein display is to produce hybrid nanoparticles in which CP fusion proteins coexist with wild-type CPs. The 2A region of the foot-and-mouth disease virus (FMDV) inserted between the protein to be displayed and the

CP allows an alternate translation of fusion and native CPs. The 2A peptide is an 18-amino-acid-long peptide containing a nonconserved region with an α-helix propensity followed by the consensus sequence Asp-(Val/Ile)-Glu-X-Asn-Pro-Gly. During translation, because of its particular structure, the 2A peptide occasionally induces a cleavage between Asn and Pro and the release of the *N*-terminal domain already synthesized. In most of the cases, translation of the remaining mRNA sequence proceeds to completion. This gives rise to a mixed population of the full length and shorter translation products (Fig. 2.13) [38, 39].

Figure 2.13 2A peptide mechanism.

Chimeric virus genomes were constructed in which the DNA sequence encoding a fusion between the green fluorescent protein (GFP) and peptide 2A was inserted in frame in front of the CP gene. A functional GFP was exposed on pepino mosaic virus (PeMV), potato virus X (PVX), and CPMV [40–42]. Using the same strategy, our group designed PVX particles displaying fully active *Candida antartica* lipase B (CalB) on their surface. The ratio between CalB-2A-CP and CPs in the hybrid VNP was 1:3. The VNPs obtained were still infectious [43].

2.2.1.2 Bacteriophages

2.2.1.2.1 *Filamentous phages and phage display*

According to their morphology, bacteriophages can be classified into filamentous (fd and M13), icosahedral (MS2 and Φ174), and prolate icosahedral phages (T4 and T7). Filamentous phages, and especially M13, have been largely employed for peptide selection [44]. The phage display technology was further used for enzyme evolution, enabling the selection of enzyme variants with improved stability, activity, and substrate specificity. For peptide screening, the peptide library is fused to the pIII protein present in five copies at one extremity of the virus. As discussed above, a direct fusion of a large protein to pIII prevents the assembly of capsid because of steric hindrance. To avoid that, phagemids have been developed. Phagemid is a M13-derived vector carrying the pIII–enzyme fusion protein, and when coexpressed in bacteria with a helper phage expressing wild-type pIII, VNP are produced, which display at their tip a single copy of the enzyme [44]. Many examples illustrating this efficient strategy may be found in the litterature. For reviews see Refs. [45–48].

More sophisticated derivatives of this technology were designed, enabling colocalization of the enzyme variants and of the selecting substrate on the same phage tip. The phosphopantetheinyl transferase (Sfp) catalyzes the transfer of various chemical groups (fluorophores, quantum dots, peptides, and nucleic acids) from their coenzyme A conjugate forms to a specific peptide substrate (ybbr) [49]. A mixed population of wild-type pIII, pIII–Sfp, and pIII–ybbr fusion protein was produced in bacteria, allowing the display of Sfp and ybbr on the same VNP tip. The versatility of this method resides in the fact that Sfp can add different groups to its ybbr substrate. Biotin ligases (Bir A) add a biotin moiety to specific AP peptidic substrates. A Bir A variant library was displayed on the phage surface together with ybbr. Sfp was used to attach an AP peptide to ybbr, and BirA variants were selected for their activity in adding biotin to this particular AP [49].

To identify peptides binding specifically to tumor receptors, phages displaying *β*-lactamase containing a peptide library fused at its *N*-terminus were produced [50–52]. *β*-lactamase served as a reporter in the screening of peptides against several tumor cell lines and purified tumor-specific receptors. Direct measurements of *β*-lactamase activity with chromogenic and fluorescent substrates avoid the step of the phage-ELISA test normally used for selection. The peptide–*β*-lactamase fusion offers a potential application in targeted enzyme-prodrug therapy.

2.2.1.2.2 *T7 phages and practical applications*

T7 phages were modified to display cellulase A (CelA), an enzyme degrading biofilms [53]. The wild-type T7 capsid contains two forms of CPs, the major coat protein 10A and the minor coat protein 10B, which accounts for 10% of the total CPs. The 10B component is longer than 10A because it is released by a translational frameshift of 10A. The frameshift of the CP was removed to produce copies of 10B fused with CelA. An additional plasmid ensuring 10A expression to allow capsid formation was introduced in the bacterium. These phages act through a "two-pronged strategy": The enzymatic activity degrades the extracellular matrix of the biofilm, exposing bacteria to the phage infection. Consequently, the phage population is amplified by bacteria. The biofilm degradation was improved when CelA was exposed on the phage surface. The type and origin of the biofilm influence the choice of the enzyme to be displayed [54, 55].

Two enzymes, xilanase A and amylase, were fused in tandem with the CP of the T7 phage. This modified T7 was tested for the treatment of paper wastes. The two enzymes were exposed in tandem at the capsid surface (capsid-Ez1-Ez2). The catalytic efficiency of the enzymes was dependent on their relative positions, that of the "in between" (i.e., capsid-Ez1-Ez2) enzyme being weaker [56].

2.2.1.2.3 *The T4 DNA packaging machine*

The T4 DNA packaging machine was modified to use the T4 VLP as a drug vector for the delivery of genes and proteins to mammalian cells [57]. Phage T4 is formed by an empty prohead, and a molecular motor composed of five subunits of the terminase protein gp17 that, driven by ATP hydrolysis, packages up to 170 kb of DNA inside the head. The head is composed of the major CP gp23, the vertex protein gp24, and the portal protein gp20. The big particularity of this virus is that its head surface is decorated by the highly antigenic outer CP Hoc (155 copies/head) and the small outer CP Soc (870/head). Soc and Hoc are not essential for the bacteriophage and allow exposition of foreign proteins fused, respectively, at the *N*- and *C*-termini of Soc and the *N*-terminus of Hoc. [58]

The assembly of the T4 phage follows a sequential procedure: prohead formation, attachment of the ATPase motor and packaging of DNA, DNA cutting and dissociation of the motor, and assembly of neck and tail proteins that seal the head. A mutant lacking neck, tail, Soc, and Hoc was developed. For example, in the wild-type T4, DNA can be packed in the head, but as this mutant is devoid of neck proteins, DNA is ejected after motor disassembly. Interestingly, the

ATPase motor can be reassembled with the empty head and used to pack nonspecific DNA materials [59]. Heads were obtaining by infecting E. coli with the mutant. After a spontaneous DNA ejection empty heads were refilled in vitro with plasmids encoding for luciferase (Luc) and GFP by the ATPase motor. Subsequently filled heads were allowed to interact with modified Soc and Hoc proteins (see Fig. 2.14). Hoc and Soc, fused, respectively, with cell-penetrating peptides (CPPs) and β-lactamase, were added to the T4-functionalized VLP. This modified T4 nanomachine is an example of a multifunctional nanocarrier enabling the in vivo delivery of both genetic material and enzymes [57].

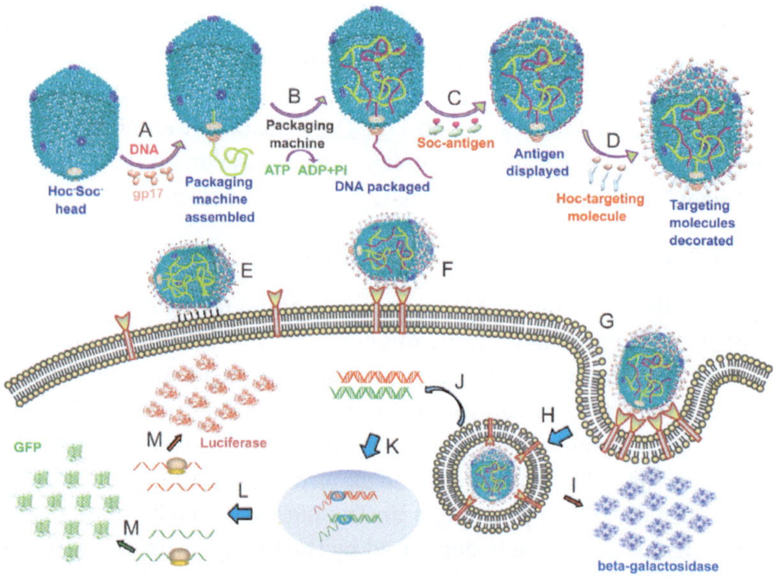

Figure 2.14 Mechanism of the T4 DNA machine for delivering proteins and genes. The Gp17 motor assembles on the Hoc⁻Soc⁻prohead (A). DNA is packaged into heads, thanks to ATP hydrolysis (B). Fusion proteins Soc-β-galactosidase (C) and Hoc-targeting molecules (CPPs) (D) associate with the head surface. The T4 machine binds to cell aspecifically (E) or by means of specific receptors (F) and is internalized (G and H). Displayed proteins (I) and genes (J, luciferase and GFP) are released in the cytosol. Genes enter the nucleus (K), followed by transcription (L) and protein expression (M). Reproduced with permission from Ref. [57].

2.2.1.2.4 *P22*

An alternative way for in vivo packaging of cargo proteins was developed for P22-derived VLPs [60]. In physiological conditions, P22 capsid formation follows a two–step process. CPs associate to form a procapsid with the help of scaffold proteins (SPs). Subsequently, the spherical prohead undergoes structural changes, releasing SPs and leading to a mature icosahedral morphology. When expressed in E. coli, CPs and SPs assemble into procapsids. Fluorescent proteins were fused to the *N*-terminus of SP interspaced by a thrombin-cleavage site. Procapsids loaded with fluorophores were then heated at 65°C, inducing the conversion in mature VLPs. SP domains were displayed on the particle surface, exposing the thrombin-cleavage site. A thrombine treatment leads to the release of SPs, with retention of the fluorescent cargo inside the VLPs. Alternatively, increasing the temperature to 75°C induces the dissociation of 12 CP pentamer vertices, resulting in a wiffleball capsid morphology and release of the cargo. Using this approach, enzymes can be confined into the particle in place of fluorophores. P22 VLPs can then work as nanoreactors, substrate access to the inside being driven by temperature modulations [60, 61].

2.2.1.3 Adenoviruses

Adenoviruses (Ads) are medium-small, nonenveloped icosahedral viruses that have a broad range of vertebrate hosts, including humans. They have been largely utilized for developing vectors for transient gene therapy as they can infect many types of cells and they don't integrate DNA in the host genome. The Ad surface has been modified in various ways for specific cell targeting. These engineering strategies could be applied for bioconjugation of proteins to a viral capsid. The adenoviral capsid is made of an assembly of three major CPs (hexon, penton, and fiber) and by three minor proteins that enhance capsid stability (pIIIa, pVIII, and pIX) [62]. Different sites of modification have been identified, allowing the display of peptides of different length. They essentially consist of the hypervariable region (HVR) of hexons, the integrin-binding RGD peptide sequence in pentons, the carboxyl terminus and the HI loop of the fiber knob protein, and the exposed carboxy termini of pIX [62, 63]. Regarding display of large proteins, it seems that pIX is the preferential site of insertion [64–67]. Although pIX is buried inside the capsid structure, GFP and enhanced yellow fluorescent protein (EYFP) were efficiently exposed on the virus surface as fusions with its *C*-terminus, which enabled tracking of Ad particles [64, 67].

It was also reported that pIX fusion to a 45 Å long linker increased the accessibility of antibodies to immunogenic peptides [68].

An improvement of this nanocarrier was obtained by functionalization of pIX with herpes simplex virus type 1 (HSV-1) thymidine kinase (TK), an enzyme largely used in cancer therapy. Analysis of cell killing after Ad infection showed that the IX–TK fusion protein could also serve as a therapeutic gene by rendering transduced cells sensitive to ganciclovir, an inhibitor of viral DNA replication [66].

A fusion protein containing TK and luciferase was inserted at the *C*-terminus of pIX interspaced by an 18-amino-acid linker containing the Flag-tag sequence (Fig. 2.15) [65]. Both enzymes retain their activity in vitro and in vivo, allowing simultaneous analysis by luminescence and positron emission tomography (PET) imaging.

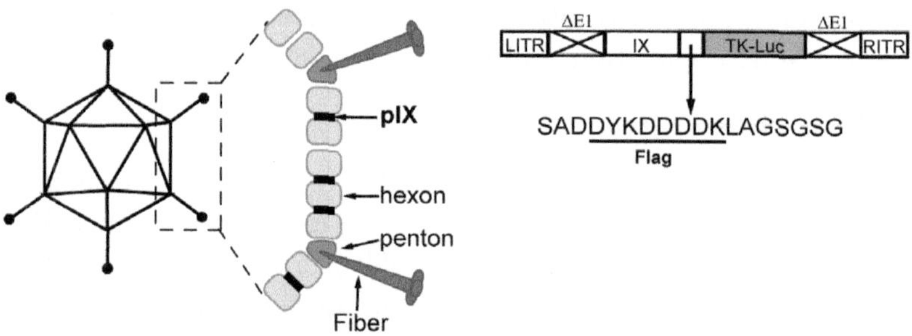

Figure 2.15 Structure of the adenoviral capsid and genetic contruct of pIX-TK-Luc.

A multifunctional Ad has been generated by replacing the native pIX gene with three expression cassettes, each one containing three different pIX fusion proteins: a Flag-tag, a His-tag, and the monomeric red fluorescent protein (mRFP) [69].

2.2.1.4 Rotaviruses

Rotaviruses are nonenveloped icosahedral viruses containing a triple capsid structure: an inner capsid (VP2), an intermediate capsid (VP6), and an outer capsid (VP7 and VP4). When coexpressed in a baculovirus–insect cell system, VP2 and VP6 can form stable VLPs. The *N*-terminus of VP2 is exposed toward the interior of the nanoparticles, and its first 92 amino acids can be removed without affecting the assembly. Two chimeric VP2 were produced in which the first 92 amino acids were replaced either by the 238-amino-acid GFP or by the 249-amino-acid DsRed protein, followed by a flexible SRGS linker [70]. Chimeric VP2, GFP-Δ92VP2, or DsRed-Δ92VP2 was then coexpressed with VP6 to obtain double-layered particles (DLPs) and with VP7 and VP4 spikes for triple-layered particles (TLPs). The complete particle GFP-Δ92VP2/6/7/4 conserves its ability to target and penetrate intestinal epithelial cells both in vitro and in vivo, including inflammatory events [71]. This system can encapsidate 120 molecules of GFP and represents a potential system for delivery of bioactive molecules into intestinal cells.

2.2.1.5 Parvoviruses

Parvoviruses are nonenveloped icosahedral viruses, and empty particles can be easily produced in a baculovirus–insect cell system. Their capsid is composed mainly of the major CP VP2 and of the minor CP VP1. It differs from VP2 by an additional 227-amino-acid *N*-terminal region pointing to the exterior. This region is not essential for particle formation, and it was substituted with lysozyme [72]. Modified VP1 was coexpressed with VP2, and each capsid contains one to two active molecules of lysozyme.

2.2.1.6 Core-like particles

Core-like particles (CLPs) are particles that can be derived from the corresponding enveloped virus.

2.2.1.6.1 *HBV CLPs and the split-core strategy*

HBV is largely employed in nanotechnology. Its core protein forms CLPs when expressed in heterologous systems. The core protein of HBV (HBc) is a 183-amino-acid protein in which the *N*-terminal 140-amino-acid domain (assembly domain) is required for assembly and the *C*-terminus is exposed to the inside of the virus and binds the DNA. The part exposed to the surface has an α-helices fold. The dimerization domain is constituted by $\alpha 3$–$\alpha 4$ that form spikes on the surface of the capsid (Fig. 2.16A). The immunogenicity of HBV resides essentially in the c/e1 loop. HBc CLPs are highly immunogenic, and the insertion of a foreign peptide in this loop has high potential for the development of vaccines [73]. Peptides of different lengths were efficiently inserted at the *N*-terminus (up to

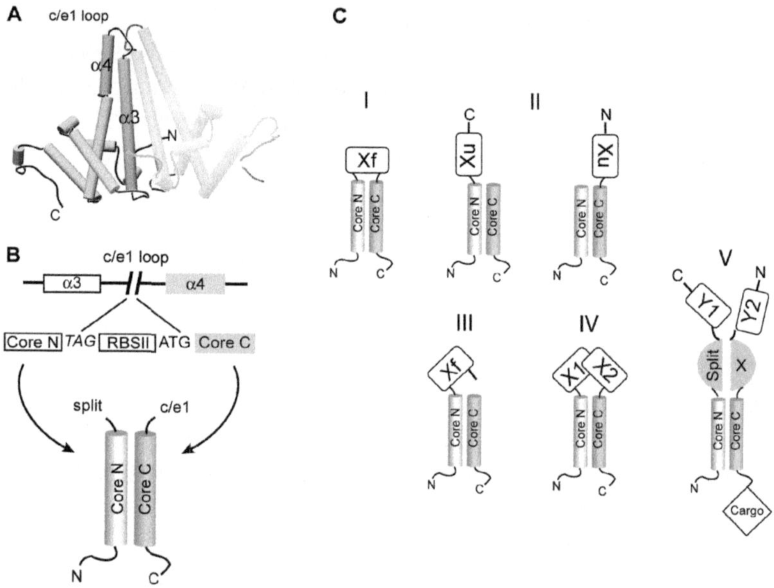

Figure 2.16 The Split-core strategy. (A) Normal configuration of the HBc assembling domain (PDB:1QGT): $\alpha3$- and $\alpha4$-helices interact and are connected by the immunogenic loop. The *C*-terminus points to the interior of the particle. (B) HBc was split in c/e1. A stop codon (left) and a start codon (right) were introduced. Core N and core C can be separately expressed in *E. coli* in a bicistronic vector with the 3′ cistron controlled by a second ribosome binding site (RBSII). (C) Different configurations that can be achieved with the Split-core strategy: favorable inserts can be directly introduced in the c/e1 loop (I), while unfavorable proteins can be fused to core N or core C (II). Even a favorable insert can be introduced in one of the two moieties of HBc, leaving one extremity for subsequent modifications (III). Two interacting domains can be separately fused to core N and core C and allowed to interact after CLP formation (IV). Finally HBc can be functionalized at different levels by introducing a targeting ligand (Y1, Y2), a monitoring reporter (split X), or a cargo fused to or in place of the *C*-terminal domain.

40-amino-acid epitopes), in the immunogenic loop (up to 100 amino acids) and at the *C*-terminus (more than 100 amino acids) [74]. Insertion in the immunodominant loop c/e1 gives the best results in term of immunogenicity. While the insertion of linear peptides is well established, direct insertion of large proteins seems to depend on their structure. Successful insertion of GFP was reported [75]. *N*- and *C*-termini of GFP are juxtaposed and insert well in acceptor sites in the HBc loop. Two different variants of the *B. burgdorferi* lipoprotein OspC, OspCa and OspCb, were successfully displayed on CLPs [76]. OspC variants share with HBc the common mushroom-shaped dimeric structure, and dimerization occurred between the two monomers exposed on the two HBc subunits.

To overcome the conformational stress imposed by the insertion of OspA in the HBc a proteolytic site was inserted at one end of the junction with the c/e1 loop [77]. A more versatile strategy was developed by Walker et al., named the SplitCore concept, in which the loop was split by inserting artificial stop and start codons (Fig. 2.16B). The *N*-terminal and *C*-terminal parts (core N and core C) can be coexpressed in *E. coli* and maintain the capacity to interact to form CLPs (Fig. 2.16B). Foreign proteins can be fused to one of the two ends, preventing conformational stress. This system is especially adapted for displaying dimeric proteins, in which the two domains can be fused separately with core N or core C. Limitations concern the two domains' solubility when expressed in *E. coli* and their interaction to reassemble the entire protein [78]. Various configurations can be obtained (see Fig. 2.16B).

To pack multiple copies of a given protein inside HBV VLPs, the *C*-terminus of HBc was fused with the 17 kDa nuclease derived from *Staphylococcus aureus*. Up to 240 nuclease subunits were packed this way [79].

2.2.2 Enveloped Viruses

Enveloped viruses are made up of a rigid proteinaceous core protecting the viral genetic materials. It is surrounded by a lipid bilayer originating from the cytoplasmic membrane of the host cell during virus budding. The lipid envelope serves as a matrix for the viral transmembrane (TM) proteins. These proteins bind to specific receptors at the cell surface and induce membrane fusion, an essential step during infection. Altering the CPs through genetic engineering may interfere with the self-assembly of the virus, whereas TM proteins are generally less sensitive to modifications.

Enveloped VLPs produced in vivo are interesting because of the possibility to incorporate complex proteins bearing the appropriate post-translational modifications in the envelope. A number of viral membrane fusion proteins are known to have permissive insertion sites for the introduction of foreign peptide sequences. When, in the membrane fusion protein, permissive sites are missing or if its engineering leads to a loss of its fusion properties, pseudotyping may give interesting results. By simultaneously infecting a cell with two different enveloped viruses, viral chimeras can be produced. The chimera particle can bring modified fusion proteins complemented by functional membrane fusion proteins [80].

2.2.2.1 Modification of viral proteins

2.2.2.1.1 *Alphaviruses*

Alphaviruses are small icosahedral enveloped viruses consisting of three concentric layers: a nucleocapsid, a host-derived lipid bilayer, and, embedded in the bilayer and protruding at the surface, 80 identical trimeric glycoproteins (spikes). While the nucleocapsid assembles independently in the cytoplasm, the polyprotein E3-E2-6K-E1 containing spikes is processed in the endoplasmic reticulum (ER) and in the Golgi apparatus. Monomeric spike proteins are addressed to the plasma membrane. The nucleocapsid core acquires its envelope during budding from the phospholipid–spike protein complex preformed in the plasma membrane [81]. Alphavirus CLPs naturally target host animal cells. They also share with some plant viruses the capacity to self-assemble in vitro in the presence of nucleic acids or other types of polianions. Moreover, if introduced in cells expressing viral glycoproteins, these CLPs can acquire an envelope. These enveloped particles, devoid of genetic material, maintain the structure of wild-type viruses and their ability to target cells and to fuse with plasma membrane to release their content in the cytoplasm. The presence of spikes in their envelope offers a point of modification to introduce new functionalities. Spikes were decorated with fluorescent proteins. Its coding sequence was inserted between

E3 and E2 into the E3-E2-6K-E1 polyprotein of Sindbis virus. Baby hamster kidney (BHK) cells were transfected with this construct. In vitro formed CLPs were introduced in transfected cells, and VLPs containing on their surface fluorescent proteins were released into the medium. Various cargos can be introduced in alphaviruses-derived VLPs. But, to ensure the right assembly of CLPs, the CP–CP interaction must not be compromised and the charge of the cargo must be negative to interact with the positive charge of the CP *N*-terminus [81,82].

2.2.2.1.2 *Retroviruses*

Retrovirus (RV)-like particles can be generated by transient transfection of common cell lines with the *gag-pol* gene [83]. RV particles begin to assemble from *gag*-precursor polyproteins anchored to the cytoplasmic membrane through myristoylation of the matrix protein (MA). During budding and after release of immature particles, processing of *gag-pol* polyproteins is initiated by the viral protease (PR) that releases single structural proteins (CA, MA, and NC are common in all RVs) and enzymes (reverse transcriptase [RT], integrase [IN]) (Fig. 2.17).

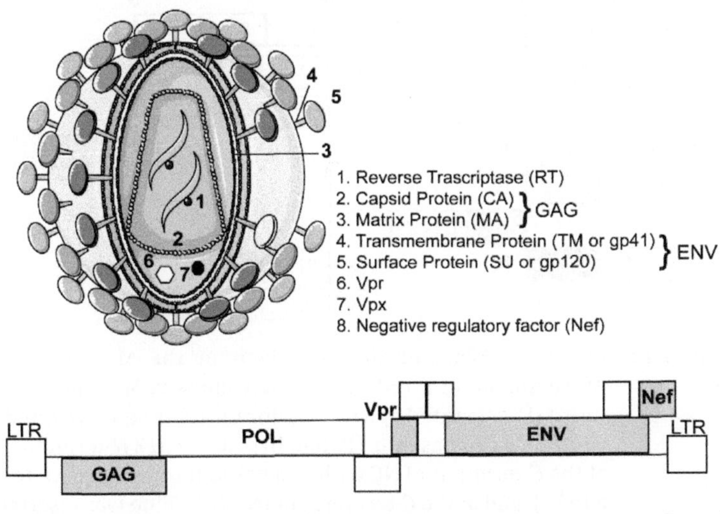

Figure 2.17 Structure and genomic organization of retroviruses. Proteins that can be modified by insertion of foreign molecules are in gray.

2.2.2.1.2.1 *Gag fusion*

To explore the possibilities of this VLP to vehiculate bioactive proteins, Voelkel et al. mapped possible insertion sites. The GFP gene was inserted in different positions in the gag-precursor of Moloney murine leukemia virus (MoMLV) (Fig. 2.18). Constructs lacking the pol genes were cotransfected with wild-type *gag-pol* [84]. Engineered particles were specifically internalized in human fibroblasts presenting the receptor. The amount of delivered proteins varied with the fusion site, with IN-GFP having weaker signal. Proteins with the nuclear signal were localized in the cell nucleus. In the various constructs, a codon-optimized version of yeast-derived Flp recombinase (Flpo) was fusioned with MA or NC with or without the proteolytic site. The enzyme activity was tested by transfection of the cells with a system switching from GFP expression to dTomato expression after Flp recombination. The enzyme retained its activity, and the signal was higher in presence of the proteolytic site. Fusion with *gag* precursors can incorporate 3000 to 5000 units of heterelogues proteins and tolerate the insertion of multiple proteolytic sites.

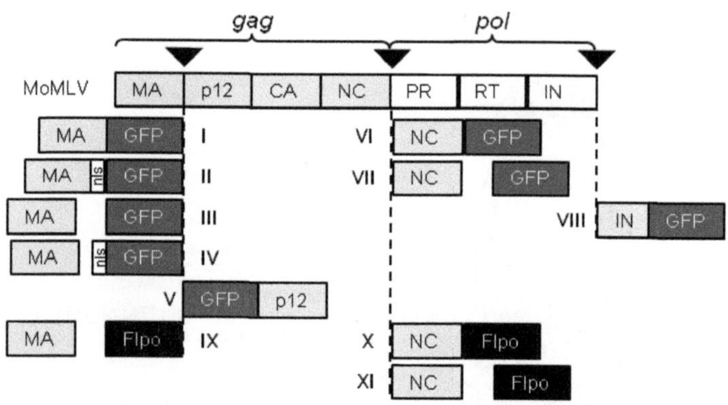

Figure 2.18 Insertion points of GFP and Flpo in the MoMLV *gag-pol* precursor. From I to IV, at the *C*-terminus of MA. Successively direct (I), separated by a nuclear localization signal, (II), and/or additional protease site (IV and III). At the p12 *N*-terminus (V), at the *C*-terminus of NC with and without the protease site (VII and VI), and at the *C*-terminus of IN (VIII). Flpo was inserted at the *C*-terminus of MA with an interposed proteolytic site (IX) and at the *C*-terminus of NC with and without the cleveage site (XI and X). Wild-type *gag-pol* was cotransfected for particle production.

This system uses *gag-pol* to engineer VLPs, but in the presence of RT and IN, the risk of integration of constructs in the host genome is still possible. It has been reported that for RV VLP formation, only the *gag* polyprotein is necessary [85–87]. Kaczmarczyk et al. generated VLPs by cotransfection of 293T competent cells with a contruct containing the *gag* gene of avian sarcoma leukosis virus (ASLV) fused with different proteins and enzymes and a vector carrying the gene encoding VSV-G, the envelope protein of vesicular stomatitis virus (VSV). Pseudotyped particles are infectious but lack the ability to replicate and to integrate the DNA of the host genome [88]. When exposed to VLPs containing fusion protein *gag*-Cre recombinase (*gag*-Cre), 293T cells containing in their genome the cassette Lox1-GFP-Lox2-RFP shift from green to red fluorescence, indicating that active Cre is present and excises the Lox1-GFP-Lox2.

In the above-mentioned example, the active enzyme was packed inside the VLPs as a fusion with the full *gag* polyprotein without the need of viral protease processing. The enzymes for which activity could be affected by *gag* fusion, cleavage is required and the viral protease must be introduced in the system for enzyme processing and release. To test how this system functions as a delivery system for enzymes converting pro-drugs to active compounds, VLPs containing *gag* fused with Fcy (cytosine deaminase) and Fur (uracil phosphorybosyltransferase) were produced and tested on cells treated with the pro-drug 5-fluorocytosine (5-FC). 5-FC is converted to 5-fluorouracil (5-FU) by Fcy and in 5-FUMP by Fur, an irreversible inhibitor acting on DNA synthesis and causing 80% reduction of the treated cells' viability. VLPs containing human caspase 8b (hCasp8B), an enzyme that causes cell death, fused with *gag* were produced too. To prevent the death of producer cells, caspase was split into two interacting domains, p10 and p18, separately fused with *gag*. Only particles containing both domains were able to cause death in the treated cells. An explanation could be that the reconstitution of caspase is guided by *gag–gag* interactions inside the particles in which the concentration is bigger than in infected cells. It was observed that influenza neuraminidase (NA) and hemagglutinin (HA) are naturally inserted in the VLP envelope during budding. Kaczmarkczyk et al. fused the cytokines' ligand both to the *C*-terminus of NA and to the *N*-terminus of HA, inducing their exposure on VLPs. This led to the death of treated tumor cells through a known receptor-dependent mechanism.

2.2.2.1.2.2 *ENV modification*

The RV envelope (*env*) protein complex is constituted of two interacting proteins, the surface (SU) protein that mediates the binding with cell receptors and the TM protein. In type C viruses like MoMLV, SU is formed by two separate globular domains, the *N*-terminal domain acting as the receptor-binding domain. The connection between the two domains is a proline-rich region (PRR) with a hypervariable *C*-terminal. This hypervariable region tolerates deletions and insertions. Large polypeptides like the 252-amino-acid single-chain antibody-binding domain (scFv) and the 96-amino-acid V1/V2 domain of human immunodeficiency virus 1 (HIV-1) can be inserted at the *C*-terminal part of the connecting domain without significant effects on structure assembly and infectivity [89]. GFP was inserted both at the *N*- and *C*-terminal part of PRR of MoMLV: only the *C*-terminal insertion gave rise to viable and infectious particles displaying GFP on their surface [90]. Avian leukosis virus (ALV) was also used as a display system for polypeptides fused with the envelope protein [91]. Polypeptides of different lengths were fused at the *N*-terminus of SU interconnected with one or more flexible linkers (G4S), enabling correct folding and oligomerization of the envelope protein and the accessibility to receptor-binding sites. The epidermal growth factor (EGF, 51 amino acids) and a single-chain antibody (α-CEAscFv, 244 amino acids) were efficiently exposed to interact with their cognate receptors. The display of foreign proteins at the envelope protein *N*-terminus of ALV particles didn't interfere with the recognition of natural receptors [91].

2.2.2.1.2.3 *Nonstructural protein modification*

- *Vpr and Vpx fusion*
  Some RVs, such as HIV-1/2, express an accessory protein Vpr, in addition to *gag-pol-env* products, that is incorporated in VPs. The function of this protein is not completely clarifyied, but it seems important in mediating nuclear transport of proviral DNA and the cell cycle [92–94]. Vpr is expressed in quantities similar to the *gag* structural protein, and it was tested for

its ability to display foreign proteins. Several features of Vpr have to be considered: a putative α-helix at the *N*-terminus of the protein, a central leucine–isoleucine region essential for packaging and nuclear transport, and a *C*-terminal arginine-rich region important for cell cycle activity. Different constructs were generated by fusing chloramphenicol acetyltranferase (CAT) at the *N*- or *C*-terminus of different Vpr-derived fragments in order to define the minimal region of Vpr for correct virion incorporation [95]. The results demonstrated that a fragment comprising amino acids 14 and 88 of HIV Vpr and containing the *N*-terminal helical region and the central leucine–isoleucine region efficiently incorporates fusion proteins into virions without disturbing the cell cycle activity. The insertion of viral protease sites between Vpr and CAT allowed the liberation of enzymes late in the virion maturation, a useful tool to prevent citotoxic enzymes from killing producer cells [95]. Structurally and functionally similar to Vpr, Vpx of HIV-2 and simian immunodeficiency virus (SIV) has been used to carry CAT and staphylococcal nuclease (SN) in VPs [96]. SN degraded the viral nucleic acids after packaging. These enzymes were fused to the *C*-terminus of full-length proteins and retained their activity after have been packaged in viruses. BlaM was inserted at the VPr *N*-terminus of HIV-1 and efficiently delivered into target cells where it catalyzed the cleavage of a fluorescent dye [97].

- *Nef fusion*

  Nef (27–34 kDa) is a myristolate regulatory protein expressed in primate lentiviruses (LVs) (HIV-1/2, SIV). The *N*-terminal myristoylation is necessary for addressing the protein to the plasma membrane and for its association with budding particles. In contrast to Vpr and Vpx, Nef is not abundant in VPs. Two Nef mutants were generated, the mutation of Gly 3 to Cys that causes a palmytoylation of the protein and a double mutant named Nef7 and containing mutations V153L and E177G. A 12-fold increase in the accumulation of the palmitoylated mutant and a 100-fold increase in the case of the double mutant Nef 7 were observed in the VLPs [98–100]. This effect was directly correlated with the mutant increase in lipid rafts (LRs) (see next paragraph), reinforcing the thesis that rafts are the sites of virus budding within the cell membrane [98, 101]. Ganciclovir (GCV) is a nucleoside analogue. Phosphorylation of GCV monophosphate by cellular kinases leads to the formation of cytotoxic GCV triphosphate, which is incorporated into DNA and inhibits DNA polymerase. HIV-1-based VLPs carrying thymidine kinase (TK) from HSV-1 fused with Nef were generated by cotransfection of competent cells with an HIV-packaging vector, a VSV-G-expressing vector, and a plasmid-containing Nef-TK construct. Nef7/TK was highly packaged in VLPs. These VLPs carrying TK induced cell death in the presence of GCV. Additional improvements were made on these VLPs by pseudotyping them with the CD4 receptor and coreceptors CXCR4 and CCR5, allowing the selective internalization of VLPs in HIV-infected cells by the "inverse fusion" process [101]. These VLPs are promising as tools for specific delivery of proteins. Proteins up to 630 amino acids could potentially be fused with Nef7 and incorporated into VLPs. In addition they may be pseudotyped with different receptors for specific targeting. Their method of preparation has been detailed [100].

2.2.2.2 Membrane anchoring

2.2.2.2.1 *GPI anchoring*

VLP budding was found associated with LRs. LRs are dynamic membrane microdomains enriched for cholesterol and sphingolipids. Some specific modifications are needed for protein association with LRs, and normally they drive the anchoring only to one side of the plasma membrane. Inner proteins are generally acylated by *N*-terminal myristoylate or palmitate, while the outer site of LRs is normally enriched in glycosylphosphatidylinositol (GPI)-linked proteins. Proteins addressed to the outer surface of LPs contain a *C*-terminal signaling sequence (GSS) that drives them to the ER where a transamidase catalyzes the substitution of the GSS domain with a preformed GPI anchor. The GPI structure is common within all GPI proteins and consists of a phosphatidylinositol group linked to proteins through a carbohydrate linker (mannose and glucosamine) [102].

The interest of LRs in VLP engineering resides in the implication of LRs in entry, assembly, and budding that have been demonstrated for several viruses (for reviews [103–107]). Additionally, it has been shown that rafts are at the basis of psedudotyping. When host cells are co-infected with two distinct viruses, their envelope proteins colocalize in these microdomains and are actively included in the viruses' envelope during budding, creating chimeric viruses with changed tropism [106, 108, 109].

The addition of a GSS signal domain to foreign proteins induces their localization into rafts. In this way, these proteins can be incorporated in VLPs during budding [106, 107]. Chimeric VLPs from SIV were decorated with a GPI-anchored granulocyte–macrophage colony-stimulating factor (GM-CSF). GM-CSF was modified by inserting the GSS signal peptide derived from the complement regulatory protein CD59 and the cell adhesion protein LFA3 and was coexpressed in insect cells with the rBV expression vector containing *gag-env* genes [110]. GM-CSF was found at the surface of VLPs and maintained its biological activity. Another form of membrane-bound GM-CSF was generated by insertion of the TM domain from the HIV *env* protein gp160. TM was able to transport GM-CSF to the plasma membrane. GM-CSF was found in the VLPs' envelope in lower amount and was less active than in VLPs containing GPI-anchored GM-CSF. It is likely that GPI-anchored proteins accumulate in rafts where prevalently virus budding takes place. Similarly several cytokines, hIL-2, hIL-4, and hGM-CSF, were fused with a GPI anchor acceptor sequence derived from Fcγ receptor III, CD16b, spaced with a polyglycine linker and associated with MoMLV-derived VLPs [111]. Again, VLPs preferably incorporated GPI-anchored cytokines instead of cytokines fused with the TM domain of type 1 TM proteins [111].

2.2.2.2.2 *Viral painting*

Metzner et al. introduced the concept of viral painting (adapted from the larger molecular painting that refers to all lipid structures) as a new postexit method to modify enveloped VLPs [112]. GPI-modified proteins are separately engineered and produced. They are subsequently reintroduced into formed VLPs after budding. The method was first described by using model GPI-linked protein

CD59 to "paint" the LV and RV vectors. Supernatants containing VPs are incubated with the recombinant CD59. Decorated particles are separated from unbound and aspecifically bound CD59 by ultracentrifugation. There were between 2 and 150 molecules (LV and RV, respectively) of CD59 per particles [113]. Two different forms of GPI-anchored GFP were produced: GPI-EH that dimerizes at high concentration and the monomeric form monoGGHis [114]. Both proteins were used to paint LV-like particles. After purification, only monomeric GFP was found on particles, while dimeric GFPs aggregated and were not incorporated in VLPs, indicating that this method is best suited for monomeric proteins. Various types of viruses were used proving the versatility of this method: LVs (HIV-1), influenza virus A, and feline herpes virus 1, all integrating the GPI-modified proteins in their envelope. Finally, two different GPI-anchored proteins, in this case, CD59 and monomeric GFP, were colocalized in viral envelopes by viral painting with a prevalence of the smaller protein [114]. The two general strategies used for the functionalization of enveloped VLPs are illustrated in Fig. 2.19.

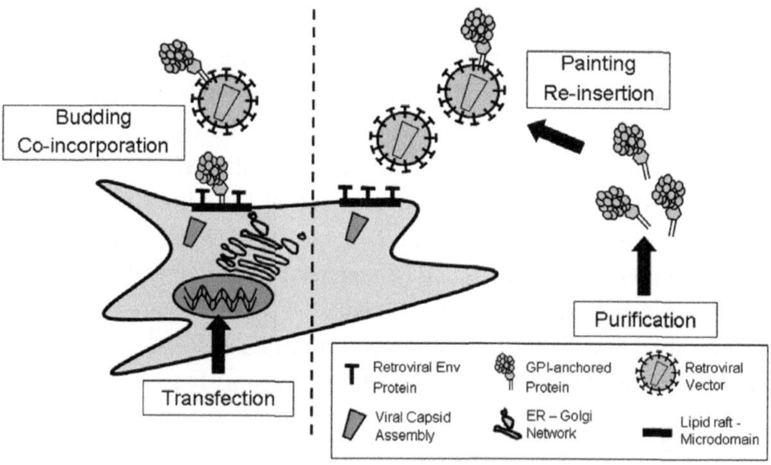

Figure 2.19 Modification of retroviral vectors with GPI-anchored proteins. GPI-anchored proteins can be inserted into the retroviral particle envelope via two mechanisms, by tranfection of retroviral producer cells (left side) or by reinsertion of purified GPI proteins, exploiting the affinity of the GPI group for the viral envelope (right side). Reproduced with permission from Ref. [107], Copyright 2008 Academic Press.

2.2.2.2.3 *Other modifications*

The envelope of HBV contains three surface proteins derived from products of the S gene: small (S, major envelope protein and also termed HBsAg, product of the entire S gene), medium (M, S + preS2 region), and large (L, S + preS2 + preS1 regions). S, M, and L proteins can be produced in yeast cells as hollow particles. They acquire their envelope during budding [115]. The presence of the human hepatocyte-recognizing peptide pre-S1 makes of L-particles a good candidate for the delivery of active biomolecules to hepatic cells [116–118]. In addition, the pre-S1 region can be easily substituted with various functionalities to change the nanoparticle target. Nucleic acids, drugs, and proteins can be introduced in L particles by electroporation [118].

Nishimura et al. focused on the fact that L particle formation starts in the ER membrane guided by accumulation of L proteins on the ER. The coexpression of L proteins with target proteins directed to the ER by lipidation enables their packaging into particles. To this purpose membrane localization sequences (MLSs), derived from *N-Ras* (an oncogene encoding for a membrane protein) and containing the CVVX motif that causes prenylation, were inserted into target proteins [119]. Two lipidation sequences were used, MLS1 (QGCMGLPCVVM), which favors both prenylation of the cysteine residue on the CVVX motif and palmytoylation of the upstream cysteine, and MLS2 (QGSMGLPCVVM), which favors only cysteine prenylation. MLSs were fused to the *C*-terminus of EGFP, and tranfection of insect cells showed their localization in the ER membrane. After cotranfection of cells with an expression vector containing the L protein–encoding sequence, target proteins were efficiently encapsidated into particles [119].

For guiding foreign proteins to influenza VP envelopes, the TM domain and cytoplasmic tail of the viral glycoproteins HA and NA were fused with target proteins, in this case avian cytokines IL2 and GM-CSF [120]. HA and NA are, respectively, type I and type II TM proteins allowing the fusion of foreign proteins either to the C- or the *N*-terminus. It has been shown that NA and HA associate with membrane rafts. This technology can be employed for decoration of any enveloped viruses and VLPs utilizing these domains during budding [121].

To enable the directed evolution of mamalian enzymes that require post-translational modifications, Granieri et al. generated an MLV-based VLP displaying on its surface the human tissue plasminogen activator (tPA). For its addressing to the cell membrane, tPA was fused to the *N*-terminus of the platelet-derived growth factor TM (PDGFRtm) domain. After cotransfection of HEK producer cells with an expression vector for *gag/pol* genes, MLV particles, each exposing a single tPA variant on its surface, were recovered. A microfluidic device was used to encapsulate the particles of MLV–tPA together with a tPA fluorescent substrate in oil–water emulsion droplets. Populations of droplets containing the tPA variants were sorted by small variations in their fluorescence. An average of 818 tPa per particle was obtained, which significantly amplified the signal [122].

2.3 Coupling of Adaptors

2.3.1 Peptides from Phage Display: The "Nanoglue" Concept

An HBV capsid-binding decapeptide (GSLLGRMKGA), also called "nanoglue," specifically interacts with spikes on HBc nanoparticles [123]. It was derived from the LLGRMK peptide initialy obtained by phage display screening of the HBV capsid [124, 125]. Nanoglue can be covalently attached to spikes by means of zero-length crosslinkers such as 1-ethyl-3-(3-dimethylaminopropyl)carbodiimide hydrochloride (EDC) and *N*-hydroxysulfosuccinimide (sulfo-NHS). The reacting lysine is crosslinked with the adjacent D64, D78, or E77 in the spike. A cell-internalizing peptide (CIP) was fused to the *N*-terminus of the nanoglue peptide, enabling its display on nanoparticles. Blokhina et al. [126, 127] engineered a chimeric CLP carrying multiple copies of the extracellular domain of the influenza virus M2 protein and interleukine 2. A chimeric protein, 3MP, was built. It was composed of a His-tagged *N*-terminus, followed by three copies of M2 derived from two different strains of the influenza virus. A nanoglue peptide was fused to its *C*-terminus. The 3MP protein and HBcs were separatly produced in *E. coli*, where the HBcs self-assembled. The lysates were then mixed, and the 3MP–HBc complex was purified by standard nickel-affinity chromatography, confirming the robustness of the nanoglue–HBc interaction. A nanoglue–interleukin 2 fusion was then displayed on 3MP–HBc to give 3MP–HBc–interleukin 2. The nanoglue concept represents a versatile method to display peptides and whole proteins on preassembled HBcs without modifing the HBV antigen. The interaction between HBcs and exposed proteins is strong enough to be easily purified by affinity chromatography without requiring covalent bioconjugation.

2.3.2 Biotin–Streptavidin

The biotin–streptavidin (or avidin) interaction (K_D = 10^{-15} mol/L) offers one of the most versatile methods of bioconjugation. VLP biotinylation can be achieved in two different ways, a nonselective direct chemical addition on amines, carboxylates, or sulfhydryls, depending on the reagent used, or the regioselective addition of a biotin on the lysine of adaptor peptide (BAP) catalyzed by the bacterial biotin ligase (BirA). Biotinylated LVs were produced by fusing BAP to the extracellular domain of the low-affinity nerve growth factor receptor (ΔLNGFR). The LNGFR–BAP fusion protein was coexpressed with BirA in HEK293T cells, where BirA catalyzes the metabolic biotinylation of BAP. Biotinylated LNGFR was addressed to the plasma membrane and incorporated in LV VLPs during budding [128]. This method was applied to baculoviruses, where BAP was fused with the glycoprotein gp164 or to the TM anchor of the VSV-G protein and to Ad in which IX protein or fiber protein was modified with BAP [129–131]. Enzymatic biotinylation can be achieved both in vitro and in vivo by coexpression of the BirA enzyme with target proteins.

In chemical biotinylation, lysine residues are the most common targets. The CP *N*-terminus of TMV was engineered with well-exposed lysines by reaction with *N*-hydroxysuccinimide (NHS) PEO_4-biotin. GFP–streptavidin (GFP–SA) fusion proteins were then allowed to react with the biotinylated virus. Abount 26% of CPs displayed GFP on the TMV surface [132].

Many examples exist of VP modification with streptavidin and avidin. Avidin was fused at the *N*-terminus of the gp64 major CP of baculovirus complemented with a wild-type copy of gp64 to ensure virus propagation. When fused to gp64, the monomeric avidin domain retained its ability to form tetrameric structures. Similar to the wild-type form of gp64, the avidin–gp64 fusion was located at the head of the virion. Avidin-modified baculoviruses were able to recognize biotinylated EGF [133]. LV particles were also pseudotyped with baculovirus gp 64 and avidin or streptavidin fused to the TM domain of the VSV protein (VSV-G EctoDomain, VSV-GED). Biotinylated ferritin and anti-EGFR antibodies were targeted toward these modified LVs [134].

2.3.3 Nucleic Acid–Guided Packaging

The MS2 phage self-assembly is guided by a 19-nucleotide packaging sequence (pac-site) carried by the viral RNA, which forms a stem loop interacting specifically with the CP. This pac-site was first used by Wu et al. as a driving force for foreign protein encapsidation into VLPs. The ricin toxin A-chain (RTA), an rRNA *N*-glycosidase that blocks the ribosomal system, was chemically crosslinked with the pac-site. Free dimeric CP units were obtained by disassembling MS2 particles. Starting from these components, VLPs containing several copies of RTA were assembled in vitro [135]. This strategy was used for encapsidation of various cargos into MS2 phages [136]. These VLPs were further functionalized for specific cell targeting as follows. CPs were modified with two peptides, SP94 recognizing specifically hepatocellular carcinoma cells and a fusogenic peptide preventing RTA degradation in the lysosome after endocytosis. RTA-loaded VLPs induce specifically apoptosis in carcinoma cells, leaving intact healthy cells [136].

This method was elegantly revisited by Finn and coworkers for the full in vivo encapsidation of functional proteins into the MS2-related Qβ bacteriophages [15, 137]. This adaptation avoids chemical conjugation. It consists of a transformation of the bacteria with a DNA vector carrying a deleted form of the virus genome. Only the CP gene was retained, and the sequence was modified as follows: An RNA aptamer that binds to the arginine-rich peptide (Rev) from the HIV-1 Tat protein was encoded on the DNA vector upstream from the ribosome-binding site (RBS); the sequence of the Qβ pac-site was inserted at the 3' end just after the CP stop codon. Different enzymes, the 25 kDa *N*-terminal aspartate dipeptidase peptidase E

(PepE), the 62 kDa firefly luciferase, and a thermostable mutant of Luc (tsLuc) were *N*-tagged with the Rev peptide and coexpressed with the modified CP in *E. coli*. Bacteria yielded VLPs loaded with foreign proteins [137]. The same strategy was also used for encapsidation of fluorescent proteins [138].

In contrast to other members of its family, like CCMV that can disassemble and reassemble in vitro into empty VLPs [139], CMV requires the presence of single-stranded DNA (ssDNA) or double-stranded DNA (dsDNA) [140]. Lu et al. demonstrated that VLP formation can be obtained with short fragments (20 nucleotides). This mechanism is not sequence dependent. To test whether the ability of DNA fragments to act as a nucleation factor can be used to carry foreign proteins at the interior of reassembled CMVs, fluorescent streptavidin was bound to a biotinylated oligonucleotide (CMV-Bt-70). VLPs containing streptavidin were obtained in the presence the biotinylated oligonucleotide [140].

2.3.4 Coiled Coils

Coiled-coil domains have been extensively used by Cornelissen's group for encapsidation of proteins and enzymes in CCMV-derived VLPs (see Chapter 3 of this book).

2.4 Immunostrategies

2.4.1 Bispecific Antibodies

Bispecific antibodies (bsMAbs) are ensuing from the combination of two different monoclonal antibodies and are able to bind two different epitopes with their two paratopes. Separate hybridomas are fused to obtain quadromas (i.e., hybrid-hybridomas) producing bifunctional antibodies.

Hybridoma producing anti-M13 antibodies were fused to an anti–alkaline phosphatase (AP) hybridoma. Quadromas producing bsMAbs were then selected for their ability to bind both M13 and AP in a one-step bridge ELISA in which the virus was coated on plates and the enzyme was in the liquid phase [141]. Bispecific antibodies recognizing both horseradish peroxidase (HRP) and the severe acute respiratory syndrome coronavirus (SARS–CoV) nucleocapsid protein N or spike protein S were developed as well [142, 143]. AP and HRP are enzymes often used in immunoassays for virus detection. Bispecific antibodies represent an improvement in diagnostics: the signal is increased, time reduced, and chemical conjugation avoided. Bispecific and monospecific antibodies have similar affinities for their viral target.

2.4.2 Z33

Our group developed a versatile system based on the interaction between three building blocks: a VP as a scaffold, a CP-directed antibody, and a fusion protein carrying the antibody-binding Z33 peptide [144]. The Z33 peptide was derived from the Staphylococcus aureus protein A (SpA). SpA is a 42 kDa protein carrying five homologous immunoglobulin-binding domains: A, B, C, D, and E. Each domain binds both the constant region (Fc) and the variable region (Fab, specifically interacting with a given antigen) of an immunoglobulin. In the immunoglobulin G (IgG) class, the Fc-binding domain is located in a turn between CH2 and CH3 (IgG's heavy chain constant domains 2 and 3). The 59-amino-acid Z domain was derived from the B domain of protein A and only binds Fc with a K_D of 10–50 nM [145]. Additional improvements of this domain led to a 33-amino-acid peptide having a similar affinity for IgGs than its wild-type counterpart but better stability [146]. Monomeric yellow fluorescent protein (mYFP) and 4-coumarate:CoA-ligase 2 (4CL2) were fused at the *C*-terminus of Z33 (Fig. 2.20B). The filamentous zucchini yellow mosaic virus (ZYMV) was used as model. Z33-tagged proteins were produced and allowed to bind in vitro ZYMV by means of CP-directed antibodies.

The assembly of the supramolecular complex was attested by correlative microscopy (transmission electron microscopy [TEM]/epifluorescence) and by affinity tests. About 87% of the viral capsid was covered. 4CL2 retains its activity when complexed with antibodies and viruses. This approach potentially allows the ordered display of any Z33-modified protein to viruses for which antibodies are available without any genetic modification of the viral genome.

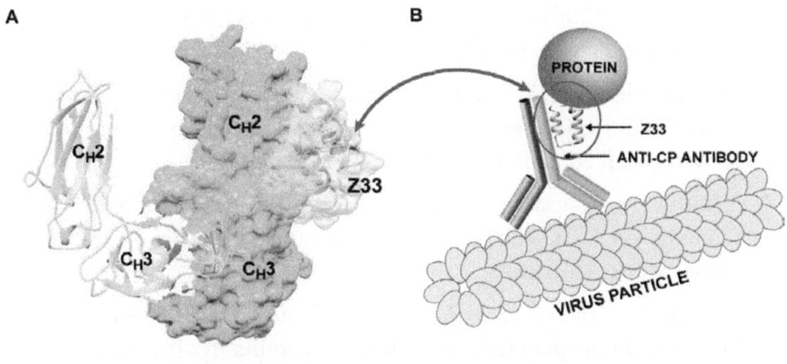

Figure 2.20 General strategy derived for coupling proteins to virus particles. (A) Interaction of Z33 with the human IgG Fc fragment. A molecular model of the Z33–Fc interaction was obtained from the complex between the human Fc fragment (cyan, surface) and fragment B of protein A13 (PDB: 1FC2). The B domain was substituted by Z33 (orange, ribbons and surface), and an energy minimization was run using Deep View (http://www.expasy.org/spdbv/). The second Fc chain (green, ribbons) of the IgG constant fragment (PDB: 1FC1) was then added to the model. (B) Scheme of the construct: the Z33 fragment (orange) genetically fused to the protein of interest binds to the Fc of an antibody raised against the virus particle. CH2 and CH3: IgG's heavy chain constant domains 2 and 3.

References

1. Stephanopoulos, N., and Francis, M. B. (2011) Choosing an effective protein bioconjugation strategy, *Nat. Chem. Biol.*, **7**(12), 876–884.
2. Trakselis, M. A., Alley, S. C., and Ishmael, F. T. (2005) Identification and mapping of protein-protein interactions by a combination of cross-linking, cleavage, and proteomics, *Bioconjugate Chem.*, **16**(4), 741–750.
3. Vasilescu, J., Guo, X., and Kast, J. (2004) Identification of protein-protein interactions using in vivo cross-linking and mass spectrometry, *Proteomics*, **4**(12), 3845–3854.
4. Song, S., Kole, S., and Bernier, M. (2012) A chemical cross-linking method for the analysis of binding partners of heat shock protein-90 in intact cells, *Biotechniques*, **0**(0), 1–7.
5. Strable, E., Prasuhn, D. E., Jr., Udit, A. K., Brown, S., Link, A. J., Ngo, J. T., Lander, G., Quispe, J., Potter, C. S., Carragher, B., Tirrell, D. A., and

Finn, M. G. (2008) Unnatural amino acid incorporation into virus-like particles, *Bioconjugate Chem.*, **19**(4), 866–875.

6. Minten, I. J., Nolte, R. J., and Cornelissen, J. J. (2010) Complex assembly behavior during the encapsulation of green fluorescent protein analogs in virus derived protein capsules, *Macromol. Biosci.*, **10**(5), 539–545.
7. Minten, I. J., Claessen, V. I., Blank, K., Rowan, A. E., Nolte, R. J. M., and Cornelissen, J. J. L. M. (2011) Catalytic capsids: the art of confinement, *Chem. Sci.*, **2**, 358–362.
8. Minten, I. J., Hendriks, L. J., Nolte, R. J., and Cornelissen, J. J. (2009) Controlled encapsulation of multiple proteins in virus capsids, *J. Am. Chem. Soc.*, **131**(49), 17771–17773.
9. Comellas-Aragonès, M., Engelkamp, H., Claessen, V. I., Sommerdijk, N. A. J. M., Rowan, A. E., Christianen, P. C. M., Maan, J. C., Verduin, B. J. M., Cornelissen, J. J. L. M., and Nolte, R. J. M. (2007) A virus-based single-enzyme nanoreactor, *Nat. Nanotechnol.*, **2**(10), 635–639.
10. Smith, M. L., Lindbo, J. A., Dillard-Telm, S., Brosio, P. M., Lasnik, A. B., McCormick, A. A., Nguyen, L. V, and Palmer, K. E. (2006) Modified tobacco mosaic virus particles as scaffolds for display of protein antigens for vaccine applications, *Virology*, **348**(2), 475–488.
11. Manchester, M., and Steinmetz, N. F. (2009) *Viruses and Nanotechnology* (Springer, Berlin, Heidelberg).
12. Steinmetz, N. F., and Manchester, M. (2011) *Viral Nanoparticles: Tools for Material Science and Biomedicine* (Pan Stanford, Singapore).
13. Bernardes, G. J. L., Chalker, J. M., Errey, J. C., and Davis, B. G. (2008) Facile conversion of cysteine and alkyl cysteines to dehydroalanine on protein surfaces: versatile and switchable access to functionalized proteins, *J. Am. Chem. Soc.*, **130**(15), 5052–5053.
14. Hermanson, G. T. (1996) *Bioconjugate Techniques* (Academic Press, San Diego, CA).
15. Rhee, J.K., Baksh, M., Nycholat, C., Paulson, J. C., Kitagishi, H., and Finn, M. G. (2012) Glycan-targeted virus-like nanoparticles for photodynamic therapy, *Biomacromolecules*, **13**(8), 2333–2338.
16. Dessens, J. T., and Lomonossoff, G. P. (1993) Cauliflower mosaic virus 35S promoter-controlled DNA copies of cowpea mosaic virus RNAs are infectious on plants, *J. Gen. Virol.*, **74**(Pt 5), 889–892.
17. Chatterji, A., Ochoa, W. F., Paine, M., Ratna, B. R., Johnson, J. E., and Lin, T. (2004) New addresses on an addressable virus nanoblock; uniquely reactive Lys residues on cowpea mosaic virus, *Chem. Biol.*, **11**(6), 855–863.

18. Chatterji, A., Ochoa, W., Shamieh, L., Salakian, S. P., Wong, S. M., Clinton, G., Ghosh, P., Lin, T., and Johnson, J. E. (2004) Chemical conjugation of heterologous proteins on the surface of cowpea mosaic virus, *Bioconjugate Chem.*, **15**(4), 807–813.

19. Brown, K. C., Yu, Z., Burlingame, A. L., and Craik, C. S. (1998) Determining protein-protein interactions by oxidative cross-linking of a glycine-glycine-histidine fusion protein, *Biochemistry*, **37**(13), 4397–4406.

20. Brown, K. C., Yang, S. H., and Kodadek, T. (1995) Highly specific oxidative cross-linking of proteins mediated by a nickel-peptide complex, *Biochemistry*, **34**(14), 4733–4739.

21. Meunier, S., Strable, E., and Finn, M. G. (2004) Crosslinking of and coupling to viral capsid proteins by tyrosine oxidation, *Chem. Biol.*, **11**(3), 319–326.

22. Michon, T., Chenu, M., Kellershon, N., Desmadril, M., and Guéguen, J. (1997) Horseradish peroxidase oxidation of tyrosine-containing peptides and their subsequent polymerization: a kinetic study, *Biochemistry*, **36**(28), 8504–8513.

23. Michon, T., Wang, W., Ferrasson, E., and Gueguen, J. (1999) Wheat prolamine crosslinking through dityrosine formation catalyzed by peroxidases: improvement in the modification of a poorly accessible substrate by "indirect" catalysis, *Biotechnol. Bioeng.*, **63**(4), 449–458.

24. Fancy, D. A., and Kodadek, T. (1998) A critical role for tyrosine residues in His6Ni-mediated protein cross-linking, *Biochem. Biophys. Res. Commun.*, **247**(2), 420–426.

25. Hein, J. E., and Fokin, V. V. (2010) Copper-catalyzed azide-alkyne cycloaddition (CuAAC) and beyond: new reactivity of copper(I) acetylides, *Chem. Soc. Rev.*, **39**(4), 1302–1315.

26. Rostovtsev, V. V, Green, L. G., Fokin, V. V, and Sharpless, K. B. (2002) A stepwise huisgen cycloaddition process: copper(I)-catalyzed regioselective "ligation" of azides and terminal alkynes, *Angew. Chem., Int. Ed. Engl.*, **41**(14), 2596–2599.

27. Tornøe, C. W., Christensen, C., and Meldal, M. (2002) Peptidotriazoles on solid phase: [1,2,3]-triazoles by regiospecific copper(i)-catalyzed 1,3-dipolar cycloadditions of terminal alkynes to azides, *J. Org. Chem.*, **67**(9), 3057–3064.

28. Wang, Q., Chan, T. R., Hilgraf, R., Fokin, V. V, Sharpless, K. B., and Finn, M. G. (2003) Bioconjugation by copper(I)-catalyzed azide-alkyne [3 + 2] cycloaddition, *J. Am. Chem. Soc.*, **125**(11), 3192–3193.

29. Sen Gupta, S., Kuzelka, J., Singh, P., Lewis, W. G., Manchester, M., and Finn, M. G. (2005) Accelerated bioorthogonal conjugation: a practical

method for the ligation of diverse functional molecules to a polyvalent virus scaffold, *Bioconjugate Chem.*, **16**(6), 1572–1579.

30. Wang, P., Tang, Y., and Tirrell, D. A. (2003) Incorporation of trifluoroisoleucine into proteins in vivo, *J. Am. Chem. Soc.*, **125**(23), 6900–6906.
31. Mock, M. L., Michon, T., van Hest, J. C. M., and Tirrell, D. A. (2006) Stereoselective incorporation of an unsaturated isoleucine analogue into a protein expressed in E. coli, *ChemBioChem*, **7**(1), 83–87.
32. Kiick, K. L., Saxon, E., Tirrell, D. A., and Bertozzi, C. R. (2002) Incorporation of azides into recombinant proteins for chemoselective modification by the Staudinger ligation, *Proc. Natl. Acad. Sci. U. S. A.*, **99**(1), 19–24.
33. Kiick, K., van Hest, J. C., and Tirrell, D. (2000) Expanding the scope of protein biosynthesis by altering the methionyl-tRNA synthetase activity of a bacterial expression host, *Angew. Chem., Int. Ed. Engl.*, **39**(12), 2148–2152.
34. Wang, L., Brock, A., Herberich, B., and Schultz, P. G. (2001) Expanding the genetic code of Escherichia coli, *Science*, **292**(5516), 498–500.
35. Canizares, M. C., Nicholson, L., and Lomonossoff, G. P. (2005) Use of viral vectors for vaccine production in plants, *Immunol. Cell Biol.*, **83**(3), 263–270.
36. Werner, S., Marillonnet, S., Hause, G., Klimyuk, V., and Gleba, Y. (2006) Immunoabsorbent nanoparticles based on a tobamovirus displaying protein A, *Proc. Natl. Acad. Sci. U. S. A.*, **103**(47), 17678–17683.
37. Kalnciema, I., Skrastina, D., Ose, V., Pumpens, P., and Zeltins, A. (2012) Potato virus Y-like particles as a new carrier for the presentation of foreign protein stretches, *Mol. Biotechnol.*, **52**(2), 129–139.
38. Ryan, M. D., King, A. M., and Thomas, G. P. (1991) Cleavage of foot-and-mouth disease virus polyprotein is mediated by residues located within a 19 amino acid sequence, *J. Gen. Virol.*, **72**(Pt 11), 2727–2732.
39. Donnelly, M. L., Luke, G., Mehrotra, A., Li, X., Hughes, L. E., Gani, D., and Ryan, M. D. (2001) Analysis of the aphthovirus 2A/2B polyprotein "cleavage" mechanism indicates not a proteolytic reaction, but a novel translational effect: a putative ribosomal "skip," *J. Gen. Virol.*, **82**(Pt 5), 1013–1025.
40. Sempere, R. N., Gómez, P., Truniger, V., and Aranda, M. A. (2011) Development of expression vectors based on pepino mosaic virus, *Plant Methods*, **7**, 6.

41. Cruz, S. S., Chapman, S., Roberts, A. G., Roberts, I. M., Prior, D. A., and Oparka, K. J. (1996) Assembly and movement of a plant virus carrying a green fluorescent protein overcoat, *Proc. Natl. Acad. Sci. U. S. A.*, **93**(13), 6286–6290.

42 Gopinath, K., Wellink, J., Porta, C., Taylor, K. M., Lomonossoff, G. P., and van Kammen, A. (2000) Engineering cowpea mosaic virus RNA-2 into a vector to express heterologous proteins in plants, *Virology*, **267**(2), 159–173.

43. Carette, N., Engelkamp, H., Akpa, E., Pierre, S. J., Cameron, N. R., Christianen, P. C., Maan, J. C., Thies, J. C., Weberskirch, R., Rowan, A. E., Nolte, R. J., Michon, T., and Van Hest, J. C. (2007) A virus-based biocatalyst, *Nat. Nanotechnol.*, **2**(4), 226–229.

44. Barbas, C. F. (2001) *Phage Display: A Laboratory Manual* (Cold Spring Harbor Laboratory Press, Cold Spring Harbor, NY).

45. Boersma, Y. L., Droge, M. J., and Quax, W. J. (2007) Selection strategies for improved biocatalysts, *FEBS J.*, **274**(9), 2181–2195.

46. Fernandez-Gacio, A., Uguen, M., and Fastrez, J. (2003) Phage display as a tool for the directed evolution of enzymes, *Trends Biotechnol.*, **21**(9), 408–414.

47. Soumillion, P., Jespers, L., Bouchet, M., Marchand-Brynaert, J., Sartiaux, P., and Fastrez, J. (1994) Phage display of enzymes and in vitro selection for catalytic activity, *Appl. Biochem. Biotechnol.*, **47**(2–3), 175–190.

48. Soumillion, P., Jespers, L., Bouchet, M., Marchand-Brynaert, J., Winter, G., and Fastrez, J. (1994) Selection of beta-lactamase on filamentous bacteriophage by catalytic activity, *J. Mol. Biol.*, **237**(4), 415–422.

49. Sunbul, M., Emerson, N., and Yin, J. (2011) Enzyme-catalyzed substrate attachment to phage surfaces for the selection of catalytic activities, *ChemBioChem*, **12**(3), 380–386.

50. Shukla, G. S., and Krag, D. N. (2010) Phage-displayed combinatorial peptide libraries in fusion to beta-lactamase as reporter for an accelerated clone screening: potential uses of selected enzyme-linked affinity reagents in downstream applications, *Com. Chem. High T. Scr.*, **13**(1), 75–87.

51. Shukla, G. S., and Krag, D. N. (2010) Cancer cell-specific internalizing ligands from phage displayed beta-lactamase-peptide fusion libraries, *Protein Eng. Des. Sel.*, **23**(6), 431–440.

52. Shukla, G. S., and Krag, D. N. (2010) Novel beta-lactamase-random peptide fusion libraries for phage display selection of cancer cell-

targeting agents suitable for enzyme prodrug therapy, *J. Drug Target,* **18**(2), 115–124.

53. Collins, J. J., Kobayashi, H., Kearn, M., Araki, M., Friedland, A., Kuan-Ta Lu, T. (2006) Engineered enzymatically active bacteriophages and methods of uses thereof (PCT/US2005/032365).

54 Lu, T. K., and Collins, J. J. (2007) Dispersing biofilms with engineered enzymatic bacteriophage, *Proc. Natl. Acad. Sci. U. S. A.,* **104**(27), 11197–11202.

55. Sutherland, I. W., Hughes, K. A., Skillman, L. C., and Tait, K. (2004) The interaction of phage and biofilms, *FEMS Microbiol. Lett.,* **232**(1), 1–6.

56. Zachariou, M., Widdowson, C. A., and Straffon, M. J. (2006) Bacteriophages displaying functional enzymes and uses thereof (PCT/AU2005/001500).

57. Tao, P., Mahalingam, M., Marasa, B. S., Zhang, Z., Chopra, A. K., and Rao, V. B. (2013) In vitro and in vivo delivery of genes and proteins using the bacteriophage T4 DNA packaging machine, *Proc. Natl. Acad. Sci. U. S. A.,* **110**(15), 5846–5851.

58. Li, Q., Shivachandra, S. B., Zhang, Z., and Rao, V. B. (2007) Assembly of the small outer capsid protein, Soc, on bacteriophage T4: a novel system for high density display of multiple large anthrax toxins and foreign proteins on phage capsid, *J. Mol. Biol.,* **370**(5), 1006–1019.

59. Zhang, Z., Kottadiel, V. I., Vafabakhsh, R., Dai, L., Chemla, Y. R., Ha, T., and Rao, V. B. (2011) A promiscuous DNA packaging machine from bacteriophage T4, *PLOS Biol.,* **9**(2), e1000592.

60. O'Neil, A., Reichhardt, C., Johnson, B., Prevelige, P. E., and Douglas, T. (2011) Genetically programmed in vivo packaging of protein cargo and its controlled release from bacteriophage P22, *Angew. Chem., Int. Ed. Engl.,* **50**(32), 7425–7428.

61. Kang, S., Uchida, M., O'Neil, A., Li, R., Prevelige, P. E., and Douglas, T. (2010) Implementation of p22 viral capsids as nanoplatforms, *Biomacromolecules,* **11**(10), 2804–2809.

62. Singh, R., and Kostarelos, K. (2009) Designer adenoviruses for nanomedicine and nanodiagnostics, *Trends Biotechnol.,* **27**(4), 220–229.

63. Campos, S. K., and Barry, M. A. (2007) Current advances and future challenges in Adenoviral vector biology and targeting, *Curr. Gene Ther.,* **7**(3), 189–204.

64. Meulenbroek, R. A., Sargent, K. L., Lunde, J., Jasmin, B. J., and Parks, R. J. (2004) Use of adenovirus protein IX (pIX) to display large

polypeptides on the virion--generation of fluorescent virus through the incorporation of pIX-GFP, *Mol. Ther.*, **9**(4), 617–624.

65. Matthews, Q. L., Sibley, D. A., Wu, H., Li, J., Stoff-Khalili, M. A., Waehler, R., Mathis, J. M., and Curiel, D. T. (2006); genetic incorporation of a herpes simplex virus type 1 thymidine kinase and firefly luciferase fusion into the adenovirus protein IX for functional display on the virion, *Mol. Imaging*, **5**(4), 510–519.
66. Li, J., Le, L., Sibley, D. A., Mathis, J. M., and Curiel, D. T. (2005) Genetic incorporation of HSV-1 thymidine kinase into the adenovirus protein IX for functional display on the virion, *Virology*, **338**(2), 247–258.
67. Zakhartchouk, A., Connors, W., van Kessel, A., and Tikoo, S. K. (2004) Bovine adenovirus type 3 containing heterologous protein in the C-terminus of minor capsid protein IX, *Virology*, **320**(2), 291–300.
68. Belousova, N., Krendelchtchikova, V., Curiel, D. T., and Krasnykh, V. (2002) Modulation of adenovirus vector tropism via incorporation of polypeptide ligands into the fiber protein, *J. Virol.*, **76**(17), 8621–8631.
69. Tang, Y., Wu, H., Ugai, H., Matthews, Q. L., and Curiel, D. T. (2009) Derivation of a triple mosaic adenovirus for cancer gene therapy, *PLOS ONE*, **4**(12), e8526.
70. Charpilienne, A., Nejmeddine, M., Berois, M., Parez, N., Neumann, E., Hewat, E., Trugnan, G., and Cohen, J. (2001) Individual rotavirus-like particles containing 120 molecules of fluorescent protein are visible in living cells, *J. Biol. Chem.*, **276**(31), 29361–29367.
71. Cortes-Perez, N. G., Sapin, C., Jaffrelo, L., Daou, S., Grill, J. P., Langella, P., Seksik, P., Beaugerie, L., Chwetzoff, S., and Trugnan, G. (2010) Rotavirus-like particles: a novel nanocarrier for the gut, *J. Biomed. Biotechnol.*, **2010**, 317545.
72. Miyamura, K., Kajigaya, S., Momoeda, M., Smith-Gill, S. J., and Young, N. S. (1994) Parvovirus particles as platforms for protein presentation, *Proc. Natl. Acad. Sci. U. S. A.*, **91**(18), 8507–8511.
73. Pumpens, P., and Grens, E. (2001) HBV core particles as a carrier for B cell/T cell epitopes, *Intervirology*, **44**(2–3), 98–114.
74. Ulrich, R., Nassal, M., Meisel, H., and Krüger, D. H. (1998) Core particles of hepatitis B virus as carrier for foreign epitopes, *Adv. Virus Res.*, **50**, 141–182.
75. Kratz, P. A., Bottcher, B., and Nassal, M. (1999) Native display of complete foreign protein domains on the surface of hepatitis B virus capsids, *Proc. Natl. Acad. Sci. U. S. A.*, **96**(5), 1915–1920.

76. Skamel, C., Ploss, M., Bottcher, B., Stehle, T., Wallich, R., Simon, M. M., and Nassal, M. (2006) Hepatitis B virus capsid-like particles can display the complete, dimeric outer surface protein C and stimulate production of protective antibody responses against Borrelia burgdorferi infection, *J. Biol. Chem.*, **281**(25), 17474–17481.

77. Walker, A., Skamel, C., Vorreiter, J., and Nassal, M. (2008) Internal core protein cleavage leaves the hepatitis B virus capsid intact and enhances its capacity for surface display of heterologous whole chain proteins, *J. Biol. Chem.*, **283**(48), 33508–33515.

78. Walker, A., Skamel, C., and Nassal, M. (2011) SplitCore: an exceptionally versatile viral nanoparticle for native whole protein display regardless of 3D structure, *Sci. Rep.*, **1**, 5.

79. Beterams, G., Böttcher, B., and Nassal, M. (2000) Packaging of up to 240 subunits of a 17 kDa nuclease into the interior of recombinant hepatitis B virus capsids, *FEBS Lett.*, **481**(2), 169–176.

80. Fischlechner, M., and Donath, E. (2007) Viruses as building blocks for materials and devices, *Angew. Chem., Int. Ed. Engl.*, **46**(18), 3184–3193.

81. Cheng, F., and Mukhopadhyay, S. (2011) Generating enveloped virus-like particles with in vitro assembled cores, *Virology*, **413**(2), 153–160.

82. Cheng, F., Tsvetkova, I. B., Khuong, Y. L., Moore, A. W., Arnold, R. J., Goicochea, N. L., Dragnea, B., and Mukhopadhyay, S. (2013) The packaging of different cargo into enveloped viral nanoparticles, *Mol. Pharm.*, **10**(1), 51–58.

83. Dalba, C., Bellier, B., Kasahara, N., and Klatzmann, D. (2007) Replication-competent vectors and empty virus-like particles: new retroviral vector designs for cancer gene therapy or vaccines, *Mol. Ther.*, **15**(3), 457–466.

84. Voelkel, C., Galla, M., Maetzig, T., Warlich, E., Kuehle, J., Zychlinski, D., Bode, J., Cantz, T., Schambach, A., and Baum, C. (2010) Protein transduction from retroviral Gag precursors, *Proc. Natl. Acad. Sci. U. S. A.*, **107**(17), 7805–7810.

85. Weldon, R. a, Erdie, C. R., Oliver, M. G., and Wills, J. W. (1990) Incorporation of chimeric gag protein into retroviral particles, *J. Virol.*, **64**(9), 4169–4179.

86. Delchambre, M., Gheysen, D., Thines, D., Thiriart, C., Jacobs, E., Verdin, E., Horth, M., Burny, A., and Bex, F. (1989) The GAG precursor of simian immunodeficiency virus assembles into virus-like particles, *EMBO J.*, **8**(9), 2653–2660.

87. Wills, J. W., Craven, R. C., and Achacoso, J. A. (1989) Creation and expression of myristylated forms of Rous sarcoma virus gag protein in mammalian cells, *J. Virol.*, **63**(10), 4331–4343.

88. Kaczmarczyk, S. J., Sitaraman, K., Young, H. A., Hughes, S. H., and Chatterjee, D. K. (2011) Protein delivery using engineered virus-like particles, *Proc. Natl. Acad. Sci. U. S. A.*, **108**(41), 16998–17003.

89. Kayman, S. C., Park, H., Saxon, M., and Pinter, A. (1999) The hypervariable domain of the murine leukemia virus surface protein tolerates large insertions and deletions, enabling development of a retroviral particle display system, *J. Virol.*, **73**(3), 1802–1808.

90. Erlwein, O., Buchholz, C. J., and Schnierle, B. S. (2003) The proline-rich region of the ecotropic Moloney murine leukaemia virus envelope protein tolerates the insertion of the green fluorescent protein and allows the generation of replication-competent virus, *J. Gen. Virol.*, **84**(Pt 2), 369–373.

91. Khare, P. D., Russell, S. J., and Federspiel, M. J. (2003) Avian leukosis virus is a versatile eukaryotic platform for polypeptide display, *Virology*, **315**(2), 303–312.

92. Rogel, M. E., Wu, L. I., and Emerman, M. (1995) The human immunodeficiency virus type 1 vpr gene prevents cell proliferation during chronic infection, *J. Virol.*, **69**(2), 882–888.

93. Heinzinger, N. K., Bukinsky, M. I., Haggerty, S. A., Ragland, A. M., Kewalramani, V., Lee, M. A., Gendelman, H. E., Ratner, L., Stevenson, M., and Emerman, M. (1994) The Vpr protein of human immunodeficiency virus type 1 influences nuclear localization of viral nucleic acids in nondividing host cells, *Proc. Natl. Acad. Sci. U. S. A.*, **91**(15), 7311–7315.

94. He, J., Choe, S., Walker, R., Di Marzio, P., Morgan, D. O., and Landau, N. R. (1995). Human immunodeficiency virus type 1 viral protein R (Vpr) arrests cells in the G2 phase of the cell cycle by inhibiting p34cdc2 activity, *J. Virol.*, **69**(11), 6705–6711.

95. Yao, X. J., Kobinger, G., Dandache, S., Rougeau, N., and Cohen, E. (1999) HIV-1 Vpr-chloramphenicol acetyltransferase fusion proteins: sequence requirement for virion incorporation and analysis of antiviral effect, *Gene Ther.*, **6**(9), 1590–1599.

96. Wu, X., Liu, H., Xiao, H., Kim, J., Seshaiah, P., Natsoulis, G., Boeke, J. D., Hahn, B. H., and Kappes, J. C. (1995) Targeting foreign proteins to human immunodeficiency virus particles via fusion with Vpr and Vpx, *J. Virol.*, **69**(6), 3389–3398.

97. Cavrois, M., De Noronha, C., and Greene, W. C. (2002) A sensitive and specific enzyme-based assay detecting HIV-1 virion fusion in primary T lymphocytes, *Nat. Biotechnol.*, **20**(11), 1151–1154.

98. Peretti, S., Schiavoni, I., Pugliese, K., and Federico, M. (2005) Cell death induced by the herpes simplex virus-1 thymidine kinase delivered by human immunodeficiency virus-1-based virus-like particles, *Mol. Ther.*, **12**(6), 1185–1196.

99. Muratori, C., D'Aloja, P., Superti, F., Tinari, A., Sol-Foulon, N., Sparacio, S., Bosch, V., Schwartz, O., and Federico, M. (2006) Generation and characterization of a stable cell population releasing fluorescent HIV-1-based virus like particles in an inducible way, *BMC Biotechnol.*, **6**(1), 52.

100. Muratori, C., Bona, R., and Federico, M. (2010) Lentivirus-based virus-like particles as a new protein delivery tool, *Methods Mol. Biol.*, **614**, 111–124.

101. Peretti, S., Schiavoni, I., Pugliese, K., and Federico, M. (2006) Selective elimination of HIV-1-infected cells by Env-directed, HIV-1-based virus-like particles, *Virology*, **345**(1), 115–126.

102. Orlean, P., and Menon, A. K. (2007) Thematic review series: lipid posttranslational modifications. GPI anchoring of protein in yeast and mammalian cells, or: how we learned to stop worrying and love glycophospholipids, *J. Lipid Res.*, **48**(5), 993–1011.

103. Chazal, N., and Gerlier, D. (2003) Virus entry, assembly, budding, and membrane rafts, *Microbiol. Mol. Biol. Rev.*, **67**(2), 226–237, table of contents.

104. Suzuki, T., and Suzuki, Y. (2006) Virus infection and lipid rafts, *Biol. Pharm. Bull.*, **29**(8), 1538–1541.

105. Suomalainen, M. (2002) Lipid rafts and assembly of enveloped viruses, *Traffic*, **3**(10), 705–709.

106. Kueng, H. J., Schmetterer, K. G., and Pickl, W. F. (2011) Lipid rafts, pseudotyping, and virus-like particles: relevance of a novel, configurable, and modular antigen-presenting platform, *Int. Arch. Allergy Immunol.*, **154**(2), 89–110.

107. Metzner, C., Salmons, B., Günzburg, W. H., and Dangerfield, J. A. (2008) Rafts, anchors and viruses: a role for glycosylphosphatidylinositol anchored proteins in the modification of enveloped viruses and viral vectors, *Virology*, **382**(2), 125–131.

108. Pickl, W. F., Pimentel-Muiños, F. X., and Seed, B. (2001) Lipid rafts and pseudotyping, *J. Virol.*, **75**(15), 7175–7183.

109. Briggs, J. A. G., Wilk, T., and Fuller, S. D. (2003) Do lipid rafts mediate virus assembly and pseudotyping? *J. Gen. Virol.*, **84**(Pt 4), 757–768.

110. Skountzou, I., Quan, F.-S., Gangadhara, S., Ye, L., Vzorov, A., Selvaraj, P., Jacob, J., Compans, R. W., and Kang, S.-M. (2007) Incorporation of glycosylphosphatidylinositol-anchored granulocyte-macrophage colony-stimulating factor or CD40 ligand enhances immunogenicity of chimeric simian immunodeficiency virus-like particles, *J. Virol.*, **81**(3), 1083–1094.

111. Kueng, H. J., Leb, V. M., Haiderer, D., Raposo, G., Thery, C., Derdak, S. V, Schmetterer, K. G., Neunkirchner, A., Sillaber, C., Seed, B., and Pickl, W. F. (2007) General strategy for decoration of enveloped viruses with functionally active lipid-modified cytokines, *J. Virol.*, **81**(16), 8666–8676.

112. Metzner, C., Kochan, F., and Dangerfield, J. A. (2013) Postexit surface engineering of retroviral/lentiviral vectors, *Biomed. Res. Int.*, **2013**, 253521.

113. Metzner, C., Mostegl, M. M., Günzburg, W. H., Salmons, B., and Dangerfield, J. A. (2008) Association of glycosylphosphatidylinositol-anchored protein with retroviral particles, *FASEB J.*, **22**(8), 2734–2739.

114. Metzner, C., Kochan, F., and Dangerfield, J. A. (2013) Fluorescence molecular painting of enveloped viruses, *Mol. Biotechnol.*, **53**(1), 9–18.

115. Kuroda, S., Otaka, S., Miyazaki, T., Nakao, M., and Fujisawa, Y. (1992) Hepatitis B virus envelope L protein particles. Synthesis and assembly in Saccharomyces cerevisiae, purification and characterization, *J. Biol. Chem.*, **267**(3), 1953–1961.

116. Jung, J., Matsuzaki, T., Tatematsu, K., Okajima, T., Tanizawa, K., and Kuroda, S. (2008) Bio-nanocapsule conjugated with liposomes for in vivo pinpoint delivery of various materials, *J. Controlled Release*, **126**(3), 255–264.

117. Yamada, T., Iwasaki, Y., Tada, H., Iwabuki, H., Chuah, M. K. L., VandenDriessche, T., Fukuda, H., Kondo, A., Ueda, M., Seno, M., Tanizawa, K., and Kuroda, S. (2003) Nanoparticles for the delivery of genes and drugs to human hepatocytes, *Nat. Biotechnol.*, **21**(8), 885–890.

118. Yamada, T., Jung, J., Seno, M., Kondo, A., Ueda, M., Tanizawa, K., and Kuroda, S. (2012). Electroporation and use of hepatitis B virus envelope L proteins as bionanocapsules, *Cold Spring Harbor Protoc.*, **2012**(6), 702–705.

119. Nishimura, Y., Shishido, T., Ishii, J., Tanaka, T., Ogino, C., and Kondo, A. (2012) Protein-encapsulated bio-nanocapsules production with ER membrane localization sequences, *J. Biotechnol.*, **157**(1), 124–129.

120. Yang, Y., Leggat, D., Herbert, A., Roberts, P. C., and Sundick, R. S. (2009) A novel method to incorporate bioactive cytokines as adjuvants on the surface of virus particles, *J. Interferon Cytokine Res.*, **29**(1), 9–22.

121. Veit, M., and Thaa, B. (2011) Association of influenza virus proteins with membrane rafts, *Adv. Virol.*, **2011**, 370606.

122. Granieri, L., Baret, J. C., Griffiths, A. D., and Merten, C. A. (2010) High-throughput screening of enzymes by retroviral display using droplet-based microfluidics, *Chem. Biol.*, **17**(3), 229–235.

123. Lee, K. W., Tey, B. T., Ho, K. L., Tejo, B. A., and Tan, W. S. (2012) Nanoglue: an alternative way to display cell-internalizing peptide at the spikes of hepatitis B virus core nanoparticles for cell-targeting delivery, *Mol. Pharm.*, **9**(9), 2415–2423.

124. Böttcher, B., Tsuji, N., Takahashi, H., Dyson, M. R., Zhao, S., Crowther, R. A., and Murray, K. (1998) Peptides that block hepatitis B virus assembly: analysis by cryomicroscopy, mutagenesis and transfection, *EMBO J.*, **17**(23), 6839–6845.

125. Dyson, M. R., and Murray, K. (1995) Selection of peptide inhibitors of interactions involved in complex protein assemblies: association of the core and surface antigens of hepatitis B virus, *Proc. Natl. Acad. Sci. U. S. A.*, **92**(6), 2194–2198.

126. Blokhina, E. A., Kupriyanov, V. V, Ravin, N. V, and Skryabin, K. G. (2013) The method of noncovalent in vitro binding of target proteins to virus-like nanoparticles formed by core antigen of hepatitis B virus, *Dokl. Biochem. Biophys.*, **448**, 52–54.

127. Blokhina, E. A., Kuprianov, V. V, Stepanova, L. A., Tsybalova, L. M., Kiselev, O. I., Ravin, N. V, and Skryabin, K. G. (2013) A molecular assembly system for presentation of antigens on the surface of HBc virus-like particles, *Virology*, **435**(2), 293–300.

128. Nesbeth, D., Williams, S. L., Chan, L., Brain, T., Slater, N. K. H., Farzaneh, F., and Darling, D. (2006) Metabolic biotinylation of lentiviral pseudotypes for scalable paramagnetic microparticle-dependent manipulation, *Mol. Ther.*, **13**(4), 814–822.

129. Kaikkonen, M. U., Viholainen, J. I., Närvänen, A., Ylä-Herttuala, S., and Airenne, K. J. (2008) Targeting and purification of metabolically biotinylated baculovirus, *Hum. Gene Ther.*, **19**(6), 589–600.

130. Campos, S. K., Parrott, M. B., and Barry, M. A. (2004) Avidin-based targeting and purification of a protein IX-modified, metabolically biotinylated adenoviral vector, *Mol. Ther.*, **9**(6), 942–954.

131. Parrott, M. B., Adams, K. E., Mercier, G. T., Mok, H., Campos, S. K., and Barry, M. A. (2003) Metabolically biotinylated adenovirus for cell targeting, ligand screening, and vector purification, *Mol. Ther.*, **8**(4), 688–700.

132. Smith, M. L., Lindbo, J. A., Dillard-Telm, S., Brosio, P. M., Lasnik, A. B., McCormick, A. A., Nguyen, L. V, and Palmer, K. E. (2006) Modified tobacco mosaic virus particles as scaffolds for display of protein antigens for vaccine applications, *Virology*, **348**(2), 475–488.

133. Räty, J. K., Airenne, K. J., Marttila, A. T., Marjomäki, V., Hytönen, V. P., Lehtolainen, P., Laitinen, O. H., Mähönen, A. J., Kulomaa, M. S., and Ylä-Herttuala, S. (2004) Enhanced gene delivery by avidin-displaying baculovirus, *Mol. Ther.*, **9**(2), 282–291.

134. Kaikkonen, M. U., Lesch, H. P., Pikkarainen, J., Räty, J. K., Vuorio, T., Huhtala, T., Taavitsainen, M., Laitinen, T., Tuunanen, P., Gröhn, O., Närvänen, A., Airenne, K. J., and Ylä-Herttuala, S. (2009) (Strept) avidin-displaying lentiviruses as versatile tools for targeting and dual imaging of gene delivery, *Gene Ther.*, **16**(7), 894–904.

135. Wu, M., Brown, W. L., and Stockley, P. G. (1995) Cell-specific delivery of bacteriophage-encapsidated ricin A chain, *Bioconjugate Chem.*, **6**(5), 587–595.

136. Ashley, C. E., Carnes, E. C., Phillips, G. K., Durfee, P. N., Buley, M. D., Lino, C. A., Padilla, D. P., Phillips, B., Carter, M. B., Willman, C. L., Brinker, C. J., Caldeira Jdo, C., Chackerian, B., Wharton, W., and Peabody, D. S. (2011) Cell-specific delivery of diverse cargos by bacteriophage MS2 virus-like particles, *ACS Nano*, **5**(7), 5729–5745.

137. Fiedler, J. D., Brown, S. D., Lau, J. L., and Finn, M. G. (2010) RNA-directed packaging of enzymes within virus-like particles, *Angew. Chem., Int. Ed. Engl*, **49**(50), 9648–9651.

138. Rhee, J.-K., Hovlid, M., Fiedler, J. D., Brown, S. D., Manzenrieder, F., Kitagishi, H., Nycholat, C., Paulson, J. C., and Finn, M. G. (2011) Colorful virus-like particles: fluorescent protein packaging by the Qβ capsid, *Biomacromolecules*, **12**(11), 3977–3981.

139. Zhao, X., Fox, J. M., Olson, N. H., Baker, T. S., and Young, M. J. (1995) In vitro assembly of cowpea chlorotic mottle virus from coat protein expressed in Escherichia coli and in vitro-transcribed viral cDNA, *Virology*, **207**(2), 486–494.

140. Lu, X., Thompson, J. R., and Perry, K. L. (2012) Encapsidation of DNA, a protein and a fluorophore into virus-like particles by the capsid protein of cucumber mosaic virus, *J. Gen. Virol.*, **93**(Pt 5), 1120–1126.

141. Liu, F., Guttikonda, S., and Suresh, M. (2003) Bispecific monoclonal antibodies against a viral and an enzyme: utilities in ultrasensitive virus ELISA and phage display technology, *J. Immunol. Methods*, **274**(1-2), 115–127.

142. Bhatnagar, P. K., Das, D., and Suresh, M. R. (2008) Sequential affinity purification of peroxidase tagged bispecific anti-SARS-CoV antibodies on phenylboronic acid agarose, *J. Chromatogr., B: Anal. Technol. Biomed. Life Sci.*, **863**(2), 235–241.

143. Sunwoo, H. H., Palaniyappan, A., Ganguly, A., Bhatnagar, P. K., Das, D., El-Kadi, A. O. S., and Suresh, M. R. (2013) Quantitative and sensitive detection of the SARS-CoV spike protein using bispecific monoclonal antibody-based enzyme-linked immunoassay, *J. Virol. Methods*, **187**(1), 72–78.

144. Pille, J., Cardinale, D., Carette, N., Di Primo, C., Besong-Ndika, J., Walter, J., Lecoq, H., van Eldijk, M. B., Smits, F. C. M., Schoffelen, S., van Hest, J. C. M., Mäkinen, K., and Michon, T. (2013) General strategy for ordered noncovalent protein assembly on well-defined nanoscaffolds, *Biomacromolecules*, **14**(12), 4351–4359.

145. Cedergren, L., Andersson, R., Jansson, B., Uhlén, M., and Nilsson, B. (1993) Mutational analysis of the interaction between staphylococcal protein A and human IgG1, *Protein Eng.*, **6**(4), 441–448.

146. Braisted, A. C., and Wells, J. A. (1996) Minimizing a binding domain from protein A, *Proc. Natl. Acad. Sci. U. S. A.*, **93**(12), 5688–5692.

Chapter 3

Viruses as Model Nanoreactors to Study Enzyme Kinetics

W. Frederik Rurup, Melissa S. T. Koay, and Jeroen J. L. M. Cornelissen
Department of Biomolecular Nanotechnology, MESA and the Institute for Nanotechnology, University of Twente, Enschede, the Netherlands
j.j.l.m.cornelissen@utwente.nl

In nature, the interior of a cell hosts a remarkable network of organelles and subcellular compartments in which many biochemical reactions take place. From membrane-based assemblies (such as the mitochondrion, Golgi apparatus, vesicles, and peroxisomes) to non-membrane-enclosed ribosomes, the compartmentalization of enzymes, substrates, and cofactors within a confined environment allows the cell to gain excellent spatial and temporal control over biochemical pathways [1–3]. This organization leads to a significant enhancement in substrate turnover and catalytic efficiency. As a consequence, there is much interest in the field of nanotechnology to develop enzyme nanoreactors that mimic the natural organization of cells and organelles. This involves, however, two key challenges, (1) the precise arrangement of enzymes in space and time and

Enzyme Nanocarriers
Edited by Daniela Cardinale and Thierry Michon
Copyright © 2016 Pan Stanford Publishing Pte. Ltd.
ISBN 978-981-4613-42-2 (Hardcover), 978-981-4613-43-9 (eBook)
www.panstanford.com

(2) mimicking of the molecular crowding effects imposed by the confined environment. While recent advances using top-down nanolithography approaches provide significant spatial and temporal control, it is increasingly evident that the confined environment plays a crucial role in optimizing catalytic efficiency [4]. For these reasons, much effort focused toward the development of enzyme nanoreactors, in which the enzymes are either tethered to the exterior of a nanosized assembly or encapsulated inside. While there are several examples of enzyme nanoreactors based on biosynthetic assemblies such as nanoparticles, nanopores, polymersomes, and vesicles [5, 6], such assemblies are often highly concentration dependent, display limited substrate selectivity, and have difficulties to precisely control the positioning of enzymes [7]. In recent years, much attention has turned toward protein-based assemblies, particularly viruses, as alternative building blocks for the assembly of model nanoreactors [7–12].

Glossary

Capsid: A protein shell that encapsulates the genetic material (RNA/DNA) of a virus. A virus capsid is formed by the self-assembly of multiple copies of capsid coat protein monomers into large supramolecular structures (typically rod-shaped assemblies or icosahedral bodies).

Coiled coil: A common structural motif in protein–protein interactions. Coiled coils are formed by tandem heptad sequence repeats that form α-helices. Two or more α-helices interact to form supercoiled structures. The heptad sequence repeats, namely *abcdefg*, display hydrophobic residues at positions *a* and *d*, whereas *e* and *g* are frequently occupied by charged residues that direct the orientation of the α-helices in the coiled-coil.

DNA aptamer: DNA oligonucleotides that specifically bind a target molecule.

Virus-based enzyme nanoreactor: Virus protein–cage assemblies that are designed to carry enzymes either inside or on their surface.

Virus-like particles (VLPs): Particles that are produced by the self-assembly of capsid proteins into large supramolecular assemblies. Such assemblies are morphologically similar to the capsid; however, since the capsid is devoid of the virus genetic material, such assemblies are noninfectious.

3.1 Introduction

Viruses are intrinsically attractive scaffolds due to their ability to self-assemble into highly stable symmetrical assemblies [13]. Composed of multiple copies of protein monomers, virus capsids can be readily modified (either chemically or genetically) and functionalized, making them highly versatile and tuneable scaffolds. While there are many examples of viruses as nanocarriers and (targeted) drug delivery systems, their potential application as enzyme nanoreactors has only recently emerged. This chapter provides an overview of the different virus-building blocks used in the development of enzyme nanoreactors and deals with the challenges one faces when studying enzymes within a confined environment [14], in particular how the classical laws of enzyme kinetics may no longer apply under such conditions and how viruses could be used as nanoreactors to model biological organelles.

3.2 Virus-Based Building Blocks for Nanoreactors

Although viruses can be categorized into different morphological classes, the most commonly used virus assemblies in nanotechnology are those of icosahedral and rod-shaped viruses (Fig. 3.1). In particular, the plant-based icosahedral cowpea mosaic virus (CPMV) and cowpea chlorotic mottle virus (CCMV) and the rod-shaped tobacco mosaic virus (TMV) are amongst the most widely used and most well-characterized viruses in nanotechnology [1, 15-27].

3.2.1 Plant-Based Viruses

For the use of viruses as enzyme nanoreactors, the porosity of the capsid shell plays an important role in the diffusion of substrates and products. In particular, one of the most attractive features would be a stable virus assembly that contains pores that can be reversibly controlled or triggered (e.g., by altering the pH and ionic strength). By tuning and controlling the pH, the gradual increase in shell porosity could be extremely useful for the selective uptake and

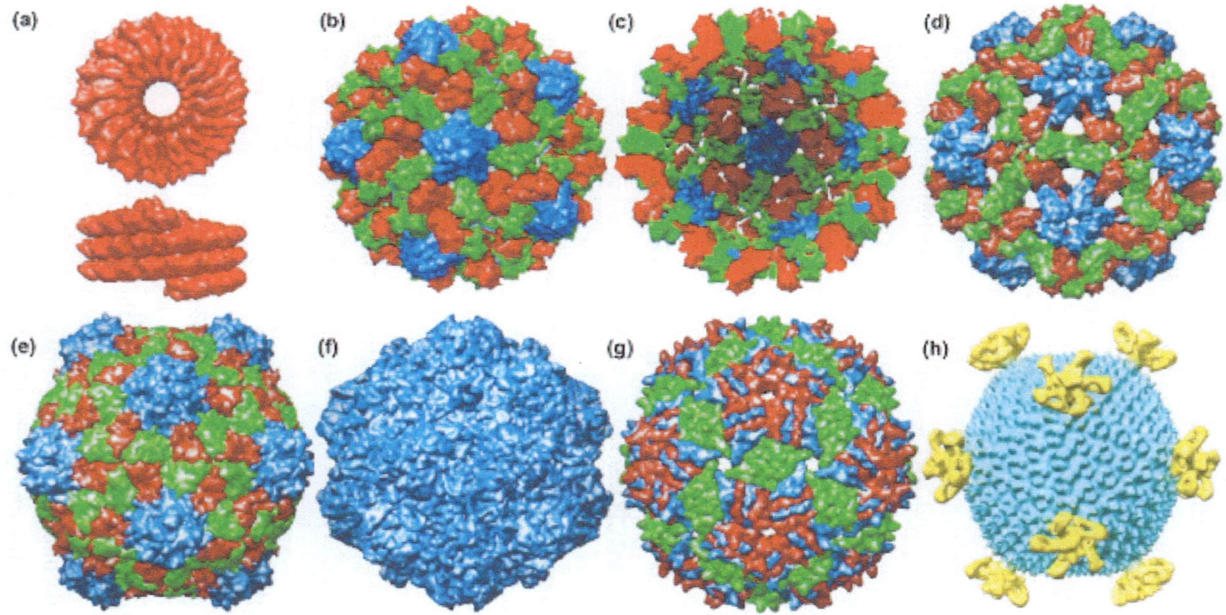

Figure 3.1 Some of the viruses currently used as model enzymes. (a) Tobacco mosaic virus (TMV), (b) cowpea chlorotic mottle virus (CCMV), (c) interior of CCMV, (d) swollen form of CCMV, (e) cowpea mosaic virus (CPMV), (f) canine parvovirus (CPV), (g) bacteriophage MS2, and (h) Sulfolobus turreted icosahedral virus (STIV). Reprinted from Ref. [28], Copyright 2010, with permission from Elsevier.

release of substrates and products. For example, CCMV is composed of 180 identical copies of a capsid coat protein that self-assembles at pH 5.0 to form the intact $T = 3$ capsid with an external diameter of 28 nm. At pH 5.0, CCMV capsids contain small <1 nm pores on the threefold axis and <0.5 nm on the fivefold axis, respectively (Fig. 3.2). At near physiological (pH ≈ 7.0), a 10% swelling occurs, while the virus assembly itself remains stable. During this process, porosity increases due to ~2 nm pores opening on the *quasi*-threefold axis, which allows diffusion of larger molecules [25]. At pH > 7.5, the CCMV capsid disassembles completely, yielding free dimers of capsid coat proteins. While the pH-driven swelling is an attractive approach to reversibly trigger changes in the porosity, gaining precise control over the pore size is extremely difficult.

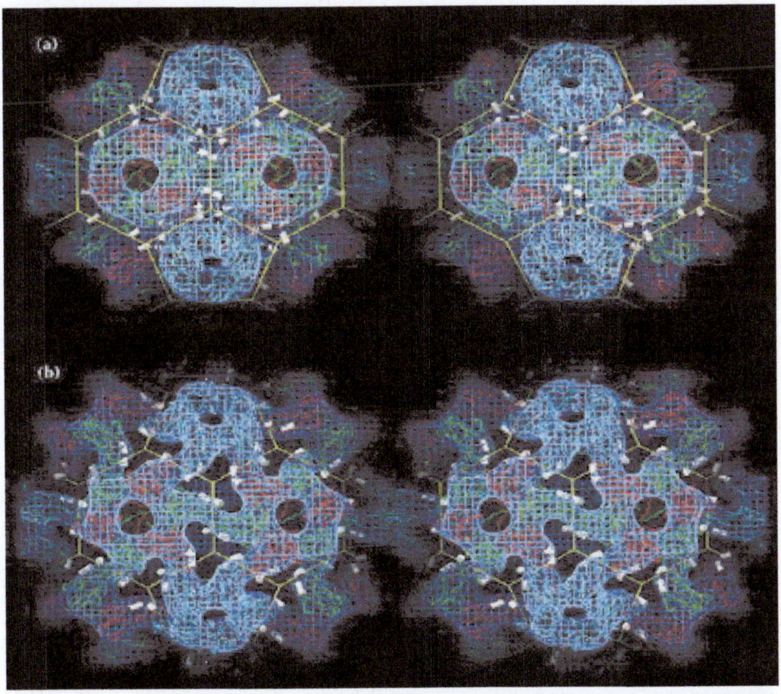

Figure 3.2 Stereoviews of cowpea chlorotic mottle virus show the capsid swelling upon increasing the pH from (a) 4.5 to (b) 7.5, resulting in an increase in the capsid pore size. Reprinted from Ref. [29], Copyright 1995, with permission from Elsevier.

3.2.2 Bacteriophages

In recent years, MS2, Q*β*, and P22 bacteriophages have been increasingly used as an alternative scaffold for virus-based assemblies. Offering intrinsically unique properties, these bacteriophages are more stable than plant-based viruses, yet they can also form empty capsid shells, making them an attractive series of virus-like nanoreactors. The native MS2 virion is composed of three components: the virus coat protein (13.7 kDa), the A protein (44.0 kDa), and a single-stranded RNA molecule. Similar to the plant-based viruses, MS2 can be reassembled from acid-driven disassembled monomers into a stable virus-like particle. Moreover, the MS2 shell possesses 32 pores per capsid, each approximately 1.6 nm in diameter, which allows the transport of potential substrates and catalysts [26]. Interestingly, Tullman-Ercek and coworkers showed that by using osmolytes such as trimethyl amine *N*-oxide (TMAO), the encapsulation efficiency of enzymes into MS2 capsids was enhanced [30]. Although osmolytes are used to regulate osmotic pressure inside cells, they have also been shown to be highly effective in stabilizing proteins [31, 32]. The use of osmolytes to alter the transport over the capsid is an interesting approach, which could be used to confer substrate selectivity or regulate the diffusion of inhibitory molecules. It is worthwhile noting that although Q*β* belongs to the same family as MS2, the pore size of Q*β* is significantly smaller than that of MS2.

P22 is a rather unique bacteriophage that is related to bacteriophage *λ*. Composed of 420 coat proteins, P22 recruits 100–330 scaffold proteins in order to assemble into the icosahedral $T = 7$ procapsid of 58 nm diameter. Upon heating to 65°C for 10 minutes, the P22 procapsid (PC) transforms to the mature capsid, whereby the capsid swells to 64 nm and the scaffold proteins are released. Interestingly, although the interior volume of the capsid is doubled in this expanded (EX) form, the pore size of 2 nm remains unaffected. Upon raising the temperature further (75°C for 20 minutes), 12 pentamers are released, irreversibly creating a wiffle ball (WB) capsid with large 10 nm sized holes (Fig. 3.3). In both the EX and WB forms, free diffusion of substrates in and out of the capsid is reported [7].

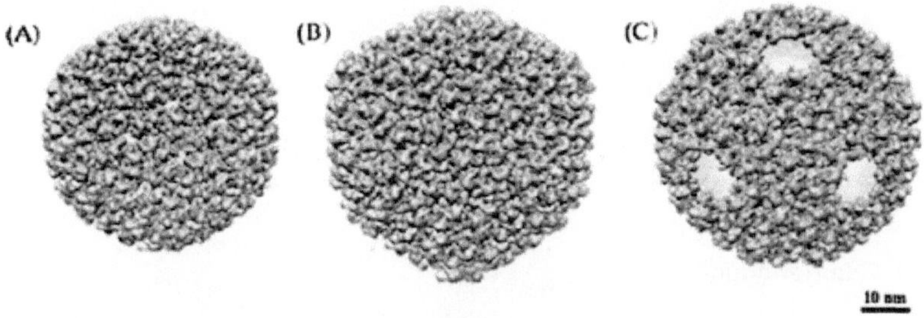

Figure 3.3 Swelling of bacteriophage P22 leads to different capsid forms. Reproduced with permission from Ref. [7]. Copyright 2012 American Chemical Society.

3.3 Enzyme-Loading Strategies

While virus-like particles form well-defined assemblies, the *controlled* loading of enzymes is a significant challenge that cannot be easily overcome. In the past, statistical encapsulation was used to load horseradish peroxidase (HRP) inside CCMV [25]. While convenient, statistical encapsulation is a highly concentration-dependent process and hence is largely uncontrollable and suffers from wide loading distributions. Since the internal walls of many viruses are highly positively charged, negatively charged linkers have been explored to promote charge–charge interactions for the controlled in vitro and in vivo loading of enzyme cargo inside virus assemblies.

3.3.1 In vitro Loading

3.3.1.1 Coiled-coil peptides

In CCMV, heterodimerizing "coiled coil" peptides were designed by fusing a positively charged α-helix to the *N*-terminus of the CCMV capsid coat protein (Fig. 3.4a). Fusion of a complementary negatively charged α-helix to the *C*-terminus of lipase B from *Pseudozyma antartica* promoted the formation of a noncovalent lipase B–CCMV complex. By varying the mixing ratios of the enzyme–capsid complex with a wild-type capsid protein the amount of loaded enzyme could be influenced upon capsid reassembly [33].

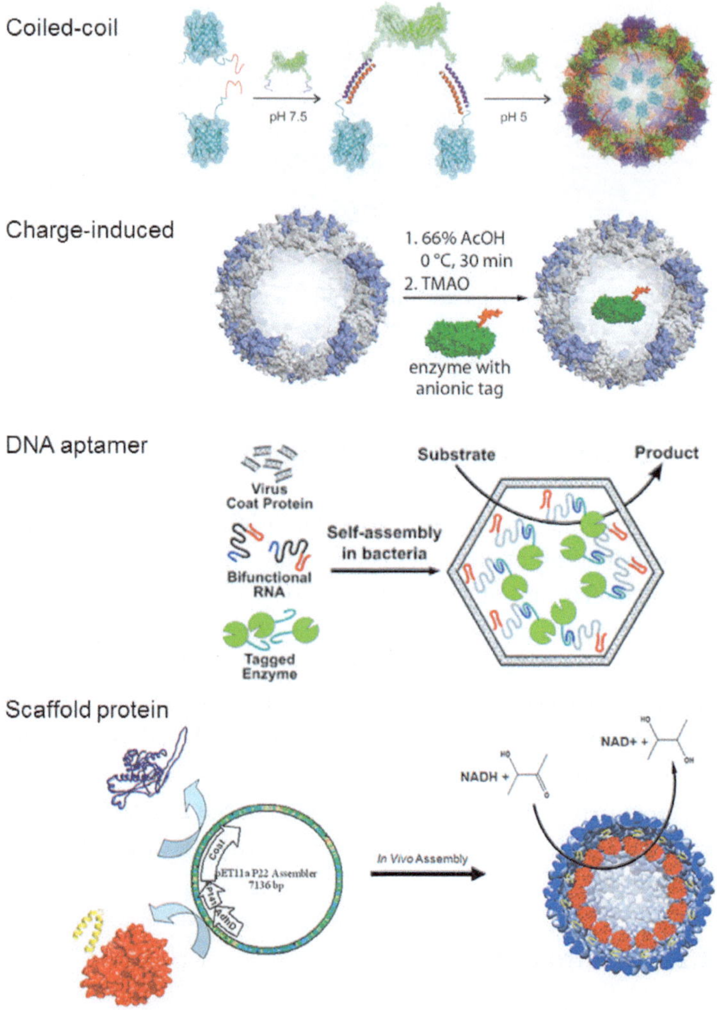

Figure 3.4 In vitro and in vivo enzyme loading strategies. Reprinted with permission from Refs. [7] and [30]. Copyright 2012 American Chemical Society.

3.3.1.2 Charge-induced loading

Inspired by the natural loading of DNA/RNA cargo, the use of nucleic acid–based linkers and their mimics has become increasingly popular to gain controlled loading. For example, MS2 bacteriophages rely on a targeting RNA hairpin for the internalization of its genomic RNA. Tullman-Ercek and coworkers recently mimicked this approach to target a non-native enzyme inside MS2 capsids [30]. In this case, a poly(16) anionic peptide linker was used as a RNA mimic, which was fused to the *C*-terminus of *E. coli* alkaline phosphatase (PhoA) to facilitate loading inside MS2 (Fig. 3.4b).

3.3.2 In vivo Loading

3.3.2.1 DNA aptamers

In some cases, the use of specific RNA/DNA recognition loops allows further control and selectivity over cargo loading. An elegant example is that of the Qβ bacteriophage, which could be loaded and assembled recombinantly under in vivo conditions. This design approach comprised two plasmids of which the first encodes expression of the virus capsid and the second, the target enzyme. The mRNA used to transcribe the capsid protein was subsequently used as an anchor. The native RNA docking sequence was engineered on the 3′ end, and on the 5′ end, an α-Rev aptamer was used. By fusing the *N*-terminal Rev-tag to the target enzyme, the two mRNA ends form "baits" for the Qβ (3′ end) and the target enzyme (5′ end). This DNA–aptamer system has already been successfully used to encapsulate peptidase E, luciferase (Luc), and a thermostable Luc variant [8], making it a versatile and effective approach for the in vivo loading of enzymes inside virus-based assemblies (Fig. 3.4c). Furthermore, by tuning the recombinant expression conditions, the amount of encapsulated enzyme can be influenced, which allows further control over enzyme loading inside the bacteriophage Qβ.

3.3.2.2 Signalling proteins

Of the viruses discussed thus far, P22 and simian virus 40 (SV40), also a $T = 7$ capsid assembly, are the only examples that rely on a scaffold protein during assembly (Fig. 3.4d). For P22, a truncated version of the scaffold protein fused to a target protein was shown to be sufficient to encapsulate the enzyme cargo during in vivo assembly. After forming the EX form, the scaffold proteins protrude out of the virus assembly, which can be subsequently removed by enzymatic digestion. Similarly, the capsid of the SV40 major capsid protein has a strong interaction with amino acids 273–308 of minor capsid protein 2. Co-infection of host cells by baculoviruses expressing either the SV40 coat protein or the cargo protein in the same cell results in stable enzyme-loaded capsids in vivo. By using their respective signaling proteins, alcohol dehydrogenase D (AdhD) from *Pyrococcus furiosus* and the homotetrameric β-glycosidase (CelB) have been successfully encapsulated in P22 and yeast cytosine deaminase (yCD) in SV40 [7, 34]. However, controlled enzyme loading has not been reported for either system so far.

3.3.3 Essential differences in loading strategies

Most encapsulation strategies are inspired on the basis of native virus loading mechanisms. To date, the different loading strategies can be classified into two broad groups, in vitro and in vivo loading. For viruses with a known assembly/disassembly pathway, in vitro loading mechanisms offer a versatile approach for controlled encapsulation of enzymes. The main benefit of in vitro loading is the possibility to purify the separate components before encapsulation. A secondary benefit is that enzyme loading can be more tightly controlled (and in some cases, predictive loading is also possible). In contrast, viruses and virus-like assemblies that undergo in vivo loading are advantageous since many of these assemblies are intrinsically more stable. However, one major disadvantage is that there is little control over the number of enzymes encapsulated, which is likely to lead to a dispersed distribution of enzyme loading between virus capsids.

3.4 Encapsulation Effects

In nature, the cell membrane provides a physical boundary to confine enzymes and substrates within a local environment. This confinement imposes a large molecular crowding effect, whereby the reduced volume increases the effective concentration of enzymes and significantly improves the kinetic efficiency of enzymes [27, 35, 36]. Similarly, the virus capsid provides a confined environment; however, the effects of encapsulating enzymes inside virus nanoreactors thus far are very diverse.

3.4.1 CCMV

The encapsulation of *Pseudozyma antartica* lipase B (PalB) inside CCMV was shown to increase the initial velocity with respect to the free enzyme in bulk solution, which may suggest that the efficiency or turnover rate of the substrate is enhanced upon encapsulation. However, interestingly, the same study also showed that encapsulation of more than one enzyme within the same capsid decreased the relative reaction rate, suggesting that, for PalB in CCMV, molecular crowding plays a significant and almost immediate role on the encapsulated enzyme. In another example, HRP was encapsulated inside CCMV. While no kinetic analyses were performed, it was shown that increasing the pH to induce capsid swelling led to increased product diffusion [37].

3.4.2 Bacteriophages MS2 and Qβ

Although the MS2 capsid is similar in size to CCMV (27 nm), it is unique as it contains 32 pores (1.6 nm in diameter) that allow diffusion of large substrates into the native capsid without altering the capsid morphology (unlike CCMV) [38]. The encapsulation of (on average) one to two PhoAs inside the MS2 bacteriophage resulted in only a marginal increase in K_m (i.e., increased apparent substrate-binding affinity) and a decrease in k_{cat} [30]. Due to the size of PhoA (94.9 kDa), Glasgow et al. attributed this decrease in enzyme efficiency to spatial crowding inside the MS2 capsid, making this a striking example of the delicate interplay between enhancement and inhibition of enzymes in confined environments.

In another example, Fiedler et al. reported encapsulation of an aspartate dipeptidase, peptidase E (PepE), in the bacteriophage Qβ [8]. In this example, the catalytic efficiency (k_{cat}/K_m) for the encapsulated PepE (2–18 enzymes per virus-like particle) increased threefold compared to free PepE. Importantly, this work showed that encapsulated PepE exhibited significantly improved thermostabilization. After heating to 50°C for 30 minutes, no loss in activity was observed for the encapsulated PepE, whereas under the same conditions, free PepE retained only 20% activity. Similarly, encapsulation was also shown to effectively protect PepE from enzymatic digestion by proteinase K (retaining 80% activity). In the same work, Luc and a thermostable luciferase (tsLuc) encapsulated in Qβ displayed an increase in K_m, suggesting that indeed the effective local substrate concentration increases inside the capsid. Interestingly, however, no enhancement of k_{cat} was observed. The authors suggested that perhaps the enzyme flexibility (or a restriction thereof) might cause this (negative) influence on kinetic performance.

3.4.3 Bacteriophage P22

The dense packing of approximately 250 alcohol dehydrogenase D (AdhD) enzymes per P22 bacteriophage led to an effective confinement molarity (M_{conf}) of 7.16 mM [7]. Significantly, Douglas and coworkers calculated that 250 AdhD led to ~27% occupancy of the total inner volume of the P22 capsid. This is a remarkable example since it closely resembles the conditions of the cell whereby the overall concentration of all enzymes and macromolecules combined is also estimated to occupy up to 30% of the physically available space [27, 35, 39, 40]. Intriguingly, the apparent k_{cat} was reported to be lower compared to the free enzyme, and the overall enzyme efficiency (k_{cat}/K_m) decreased. Similar observations were reported for encapsulation of the homotetrameric CelB [11]. Encapsulation of 84–87 CelB monomers (21 homotetramers) showed no significant change in efficiency (k_{cat}/K_m).

In general, there appears to be a trend between the virus size and the enzyme-loading number with its overall efficiency as an enzyme nanoreactor (Table 3.1). Smaller viruses such as CCMV, Qβ, and MS2, which are in the order of 24–28 nm in diameter, experience enhanced efficiency when fewer enzymes are encapsulated. Minten et al. showed that encapsulation of more than one PalB in the same CCMV capsid decreased the overall reaction rate [37]. This suggests that molecular crowding plays a significant and almost immediate role in the encapsulated enzymes in such virus assemblies. This confinement effect shows that one enzyme is enough and more just lead to steric crowding in the interior of the smaller viruses.

In contrast, the number of enzymes encapsulated in larger virus assemblies such as P22 is typically high (85–250 enzymes per capsid). However, despite the absolute number of encapsulated enzymes, these conditions closely resemble true molecular crowding and volume occupancy found in cells. Surprisingly, in both cases, the overall efficiency and the apparent affinity were reported to be lower under these conditions compared to the free enzyme. Nevertheless, it should be noted that, unlike the cell, whereby the actual concentration of a single enzyme in a cell is typically low, the encapsulation of a high number of the same enzyme might lead to self-inhibition.

Table 3.1 Overview of the confined kinetic effects observed in some model nanoreactors

Virus	Pore size	Enzyme (E)	Substrate (S)	#E	Kinetic effects
CCMV	<1 nm	PalB	DiFMU	1.3	Fivefold increase
	<1 nm	PalB	DiFMU	4.0	No change to free *E*
	<1 nm	HRP	Rhodamine	1.0	3 orders magnitude slower diffusion
	2 nm	HRP	Rhodamine	1.0	2–3 times faster diffusion over capsid
MS2	1.8 nm	PhoA	PNPP	1.6	Slight decrease
$Q\beta$	1.4 nm	PepE	Asp-AMC	9.0	No change
		Luc	Luciferin	4.0	Lower k_{cat}, higher K_m
		tsLuc	Luciferin	2.0	Equal k_{cat}, higher K_m
		tsLuc	Luciferin	9.0	Lower k_{cat}, higher K_m
P22 (PC)	2 nm	AdhD	Acetoin	250	Lower k_{cat}, lower K_m
	2 nm	CelB	PNPG	85	No change
P22 (EX)	2 nm	AdhD	Acetoin	250	Same k_{cat}, higher K_m
	2 nm	CelB	PNPG	85	No change
P22 (WB)	10 nm	AdhD	Acetoin	250	No change
	10 nm	CelB	PNPG	85	No change

3.5 Substrate Diffusion

The virus capsid not only provides a confined environment but also creates a physical barrier, which may influence the diffusion of substrates across the capsid shell and will behave different compared to traditional membrane-based organelles. There are currently few examples reporting the substrate diffusion rates between different viruses. Cornellas-Aragones et al. showed that the diffusion of rhodamine 6G in and out of CCMV is hindered for smaller pore sizes, whereas for larger pore sizes (2 nm) free diffusion was observed [25]. Recently, Douglas and coworkers reported the same observation for the bacteriophage P22 [7]. In this work, it was shown that increasing the pore size from 2 nm to 10 nm did not result in an increase in substrate and cofactor diffusion, suggesting that a pore size of 2 nm is sufficient to allow free diffusion of most substrates and products across the capsid.

3.6 Future Outlook

It has been shown that substrate diffusion, enzyme folding (and hence apparent substrate-binding affinity, K_m), and enzyme immobilization play important roles in the overall enzyme efficiency (K_{on}/K_{off} rates). Similarly, the current approaches to encapsulate enzymes inside viruses typically involve chemical or genetic modification or tethering of the enzyme, which may not only restrict enzyme activity but also restrict the rotation and conformation of the enzyme itself. Currently, all models have thus far assumed normal Michaelis–Menten enzyme kinetics. However, it is becoming increasingly apparent that standard Michaelis–Menten kinetics and, hence, the standard assumptions may not be valid for enzymes that are within a confined environment. In particular, two parameters, namely, the "free diffusion" assumption and the apparent affinity (K_{on}/K_{off} rates) need to be re-evaluated for confined systems. While fractal kinetic theory has been used to describe enzyme kinetics within the confined environment of cells and organelles, the exact parameters that define such models and their relevance to biochemical pathways are still currently under debate. In the meantime, alternative enzyme-loading strategies that do not alter or modify the enzyme itself, or reversible linkers (i.e., photocleavable linkers), could be further explored.

Nevertheless, in recent years, more virus-like assemblies such as encapsulins, carboxysomes, and de novo nanocontainers (such as lumazine synthase from *Aquifex aeolicus*) have emerged as highly promising examples for the future generation of enzyme nanoreactors. Nonetheless, the examples that currently exist for virus nanoreactors have already contributed greatly toward understanding the confined environment of organelles and cells. The enhanced efficiency of enzymes, combined with the protective properties seen in the different systems described above, still only hints at the possibilities for future enzyme nanoreactors.

Acknowledgments

This work is financially supported by a EURYI Award from the European Science Foundation (ESF)/Netherlands Organization for Scientific Research (NWO).

References

1. Cardinale, D., Carette, N., and Michon, T. (2012) Virus scaffolds as enzyme nano-carriers, *Trends Biotechnol.*, **30**(7), 369–376.
2. Tanaka, S., Sawaya, M. R., and Yeates, T. O. (2010) Structure and mechanisms of a proteinbased organelle in Escherichia coli, *Science*, **327**(5961), 81–84.
3. Yeates, T. O., Crowley, C. S., and Tanaka, S. (2010) Bacterial microcompartment organelles: protein shell structure and evolution, *Annu. Rev. Biophys.*, **39**, 185–205.
4. O'Neil, A., Reichhardt, C., Johnson, B., Prevelige, P. E., and Douglas, T. (2011) Genetically programmed in vivo packaging of protein cargo and its controlled release from bacteriophage P22, *Angew. Chem., Int. Ed.*, **50**(32), 7425–7428.
5. Tanner, P., Onaca, O., Balasubramanian, V., Meier, W., and Palivan, C. G. (2011) Enzymatic cascade reactions inside polymeric nanocontainers: a means to combat oxidative stress, *Chem. Eur. J.*, **17**(16), 4552–4560.
6. van Dongen, S. F. M., Nallani, M., Cornelissen, J. J. L. M., Nolte, R. J. M., and van Hest, J. C. M. (2009) A three-enzyme cascade reaction through positional assembly of enzymes in a polymersome nanoreactor, *Chem. Eur. J.*, **15**(5), 1107–1114.
7. Patterson, D. P., Prevelige, P. E., and Douglas, T. (2012) Nanoreactors by programmed enzyme encapsulation inside the capsid of the bacteriophage P22, *ACS Nano*, **6**, 5000–5009.

8. Fiedler, J. D., Brown, S. D., Lau, J. L., and Finn, M. G. (2010) RNA-directed packaging of enzymes within virus-like particles, *Angew. Chem., Int. Ed.*, **49**(50), 9648–9651.
9. Lee, L., Niu, Z., and Wang, Q. (2009) Viruses and virus–like protein assemblies: chemically programmable nanoscale building blocks, *Nano Res.*, **2**(5), 349–364.
10. Minten, I. J., Hendriks, L. J. A., Nolte, R. J. M., and Cornelissen, J. J. L. M. (2009) Controlled encapsulation of multiple proteins in virus capsids, *J. Am. Chem. Soc.*, **131**(49), 17771–17773.
11. Patterson, D. P., Schwarz, B., El-Boubbou, K., Oost, J. van der, Prevelige, P. E., and Douglas, T. (2012) Virus-like particle nanoreactors: programmed encapsulation of the thermostable CelB glycosidase inside the P22 capsid, *Soft Matter*, **8**(39), 10158–10166.
12. Worsdorfer, B., Woycechowsky, K. J., and Hilvert, D. (2011) Directed evolution of a protein container, *Science*, **331**(6017), 589–592.
13. Arora, P. S., and Kirshenbaum, K. (2004) Nano-tailoring: stitching alterations on viral coats, *Chem. Biol.*, **11**(4), 418–420.
14. Hou, C., Luo, Q., Liu, J., Miao, L., Zhang, C., Gao, Y., Zhang, X., Xu, J., Dong, Z., and Liu, J. (2012) Construction of GPx active centers on natural proteinnanodisk/nanotube: a new way to develop artificial nanoenzyme, *ACS Nano*, **6**(10), 8692–8701.
15. Aljabali, A. A., Barclay, J. E., Steinmetz, N. F., Lomonossoff, G. P., and Evans, D. J. (2012) Controlled immobilisation of active enzymes on the cowpea mosaic virus capsid, *Nanoscale*, **4**(18), 5640–5645.
16. Destito, G., Schneemann, A., and Manchester, M. (2009) Biomedical nanotechnology using virus-based nanoparticles, *Curr. Top. Microbiol. Immunol.*, **327**, 95–122.
17. Manchester, M., and Singh, P. (2006) Virus-based nanoparticles (VNPs): platform technologies for diagnostic imaging, *Adv. Drug Delivery Rev.*, **58**(14), 1505–1522.
18. Steinmetz, N. F., Lin, T., Lomonossoff, G. P., and Johnson, J. E. (2009) Structure-based engineering of an icosahedral virus for nanomedicine and nanotechnology, *Curr. Top. Microbiol. Immunol.*, **327**, 23–58.
19. Bruckman, M. A., Soto, C. M., McDowell, H., Liu, J. L., Ratna, B. R., Korpany, K. V., Zahr, O. K., and Blum, A. S. (2011) Role of hexahistidine in directed nanoassemblies of tobacco mosaic virus coat protein, *ACS Nano*, **5**(3), 1606–1616.

20. Gerasopoulos, K., McCarthy, M., Banerjee, P., Fan, X., Culver, J. N., and Ghodssi, R. (2010) Biofabrication methods for the patterned assembly and synthesis of viral nanotemplates, *Nanotechnology*, **21**(5), 055304.
21. Liu, N., Wang, C., Zhang, W., Luo, Z., Tian, D., Zhai, N., Zhang, H., Li, Z., Jiang, X., Tang, G., and Hu, Q. (2012) Au nanocrystals grown on a better-defined one-dimensional tobacco mosaic virus coated protein template genetically modified by a hexahistidine tag, *Nanotechnology*, **23**(33), 335602.
22. Wen, A. M., Lee, K. L., Yildiz, I., Bruckman, M. A., Shukla, S., and Steinmetz, N. F. (2012) Viral nanoparticles for in vivo tumor imaging, *J. Vis. Exp.*, **69**, e4352–e4352.
23. Soto, C. M., and Ratna, B. R. (2010) Virus hybrids as nanomaterials for biotechnology, *Curr. Opin. Biotechnol.*, **2**(4), 426–438.
24. Carette, N., Engelkamp, H., Akpa, E., Pierre, S. J., Cameron, N. R., Christianen, P. C. M., Maan, J. C., Thies, J. C., Weberskirch, R., Rowan, A.E., Nolte, R. J. M., Michon, T., and Van Hest, J. C. M. (2007) A virus-based biocatalyst, *Nat. Nanotechnol.*, **2**(4), 226–229.
25. Cornellas-Aragones, M., Engelkamp, H., Claessen, V. I., Sommerdijk, N.A., Rowan, A. E., Christianen, P. C., Maan, J. C., Verduin, B. J., Cornelissen, J. J., and Nolte, R. J. (2007) A virus-based single-enzyme nanoreactor, *Nat. Nanotechnol.*, **2**(10), 635–639.
26. de la Escosura, A., Nolte, R. J. M., and Cornelissen, J. J. L. M. (2009) Viruses and protein cages as nanocontainers and nanoreactors, *J. Mater. Chem.*, **19**(16), 2274–2278.
27. Dunker, A. K., and Fernandez, A. (2007) Engineering productive enzyme confinement, *Trends Biotechnol.*, **25**(5), 189–190.
28. Koudelka, K. J., and Manchester, M. (2010) Chemically modified viruses: principles and applications, *Curr. Opin. Chem. Biol.*, **14**(6), 810–817.
29. Speir, J. A., Munshi, S., Wang, G., Baker, T., and Johnson, J. E. (1995) Structures of the native and swollen forms of cowpea chlorotic mottle virus determined by X-ray crystallography and cryo-electron microscopy, *Structure*, **3**(1), 63–78.
30. Glasgow, J. E., Capehart, S. L., Francis, M. B., and Tullman-Ercek, D. (2012) Osmolyte-mediated encapsulation of proteins inside MS2 viral capsids, *ACS Nano*, **6**(10), 8658–8664.
31. Auton, M., Roesgen, J., Sinev, M., Holthauzen, L. M. F., and Bolen, D. W. (2011) Osmolyte effects on protein stability and solubility: a balancing act between backbone and side-chains, *Biophys. Chem.*, **159**(1), 90–99.

32. Khan, S. H., Ahmad, N., Ahmad, F., and Kumar, R. (2010) Naturally occurring organic osmolytes: from cell physiology to disease prevention, *Iubmb Life*, **62**(12), 891–895.

33. Minten, I. J., Nolte, R. J. M., and Cornelissen, J. J. L. M. (2010) Complex assembly behavior during the encapsulation of green fluorescent protein analogs in virus derived protein capsules, *Macromol. Biosci.*, **10**(5), 539–545.

34. Inoue, T., Kawano, M. A., Takahashi, R. U., Tsukarnoto, H., Enornoto, T., Imai, T., Kataoka, K., and Handa, H. (2008) Engineering of SV40-based nano-capsules for delivery of heterologous proteins as fusions with the minor capsid proteins VP2/3, *J. Biotechnol.*, **134**(1–2), 181–192.

35. Ellis, R. J. (2001) Macromolecular crowding: obvious but underappreciated, *Trends Biochem. Sci.*, **26**(10), 597–604.

36. Zhou, H. X., Rivas, G. N., and Minton, A. P. (2008) Macromolecular crowding and confinement: biochemical, biophysical, and potential physiological consequences, *Annu. Rev. Biophys.*, **37**, 375–397.

37. Minten, I. J., Claessen, V. I., Blank, K., Rowan, A. E., Nolte, R. J. M., and Cornelissen, J. J. L. M. (2011) Catalytic capsids: the art of confinement, *Chem. Sci.*, **2**(2), 358–362.

38. Witus, L. S., and Francis, M. B. (2011) Using Synthetically modified proteins to make new materials, *Acc. Chem. Res.*, **44**(9), 774–783.

39. Ellis, R. J. (2001) Macromolecular crowding: an important but neglected aspect of the intracellular environment, *Curr. Opin. Struct. Biol.*, **11**(1), 114–119.

40. Minton, A. P. (2001) The influence of macromolecular crowding and macromolecular confinement on biochemical reactions in physiological media, *J. Biol. Chem.*, **276**(14), 10577–10580.

Chapter 4

Nanoscale Compartmentalization Techniques in Cascade Catalysis

Jorgen S. Willemsen, Floris P. J. T. Rutjes, and Jan C. M. van Hest
Radboud University Nijmegen, Institute for Molecules and Materials, Heyendaalseweg 135, 6525 AJ Nijmegen, the Netherlands
j.vanhest@science.ru.nl

4.1 Cascade Reactions and Incompatibility Issues

The highly efficient formation of complex molecules in nature can be attributed to sophisticated, well-developed reaction sequences. In living cells, production of molecules often proceeds via multiple-step cascade reactions. A precise control of the reaction sequence is achieved by positional assembly or spatial allocation of the responsible enzymes. In addition, this strategy suppresses incompatibility issues between the different reactions [1, 2]. Well-known cyclic reaction sequences in nature are the citric acid cycle [3] and the Calvin cycle [4]. Examples of complex linear sequences are fatty acid oxidation [5], the production of vitamin C [6], and taxol synthesis [7], the

Enzyme Nanocarriers
Edited by Daniela Cardinale and Thierry Michon
Copyright © 2016 Pan Stanford Publishing Pte. Ltd.
ISBN 978-981-4613-42-2 (Hardcover), 978-981-4613-43-9 (eBook)
www.panstanford.com

latter being a highly active natural product that is currently used as a medicine. The transformations are often enantioselective, and the occurrence of undesired side reactions or the formation of waste products is negligible. Transport of labile reaction intermediates occurs sometimes via channels that connect two active sites, called metabolic channeling [8]. Remarkably, the separate enzymatic steps do not hamper each other in an undesired way, and the biocatalysts retain their activity for a prolonged period of time.

A related strategy in nature to prevent harmful interactions between catalytic systems is by compartmentalization. Processes can be separated from each other using different membrane-based compartments for the formation of intermediate structures. Substrates and products of reactions can easily move between compartments, sometimes via selective membrane channels [9]. Another feature of this compartmentalization strategy is that all enzymatic reactions that are in equilibrium are pushed to one site by the in situ product removal strategy: products are continuously removed from a compartment [10]. In situ product removal is also beneficial if product inhibition of enzymes occurs.

On the basis of the efficient results obtained in nature, chemists have applied this knowledge to construct one-pot reaction sequences in vitro [11–14]. In such cascade processes, steps A→B, B→C, C→D, and so on happen concurrently so that only one purification step is necessary. One-pot reactions are not always cascade reactions; however, reactions in one pot can also be performed subsequently. In other words, some reagents of a next reaction are added after the completion of a previous reaction. This means that the reactions do not proceed simultaneously. Replacing multistep batch-wise processes by one-pot procedures increases efficiency, as fewer purification steps of intermediate compounds are necessary. In a batch-wise process, time-consuming and waste-producing purifications are performed between individual reactions and multiple purification steps are usually required (Fig. 4.1). Another advantage is that reactive compounds are generated in situ and directly reacted into another compound, thereby preventing side reactions. Considering the strong points of cascade chemistry, it is no surprise that it has been proposed to contribute to greener chemical processes [15] and that it has continuously received attention of the chemical community [16].

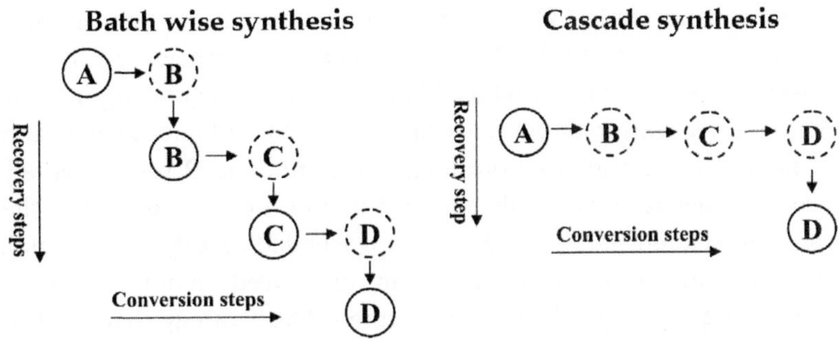

Figure 4.1 Schematic difference between step-wise and cascade synthesis. Reproduced with permission from Ref. [11]. Copyright (2003) American Chemical Society.

The major challenge in combining multiple chemical transformations is the compatibility of different reagents and catalysts. One approach for a successful reaction sequence is to search for two or more compatible reactions. Thus, none of the chemicals present in the mixture inhibit another reaction in the sequence. Many combinations of compatible reactions have been reported in the literature [11–14]. In a remarkable example described by Orru, an eight-component one-pot reaction was carried out with a total reaction yield of 24% (Fig. 4.2) [17].

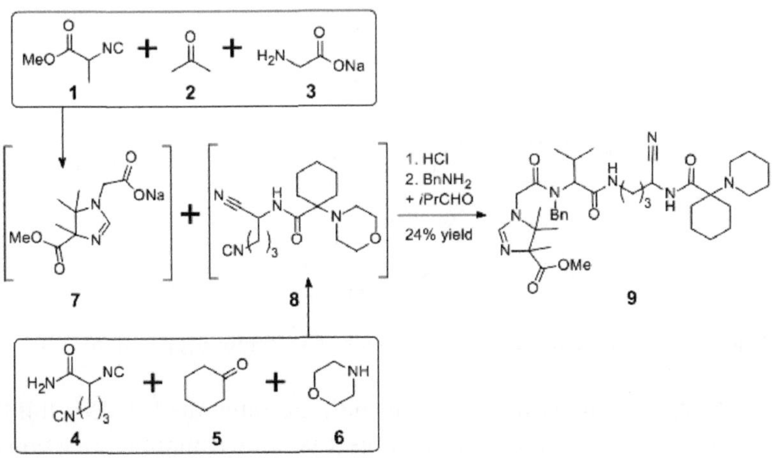

Figure 4.2 Eight-component one-pot reaction [17].

First, isocyanide (**1**) reacted with acetone (**2**) and amino acid (**3**). After the formation of **7**, compounds **4**, **5**, and **6** were added to the same reaction vessel, which led to the formation of compound **8**. After neutralization of the reaction mixture by HCl and the addition of benzylamine and isobutyraldehyde, compound **9** was obtained. Product formation was only successful after a careful design of the reaction sequence. All the reactions proceeded readily in methanol, intermediates did not react in an undesired manner, and no interfering waste products were formed. The strategy that worked here was to perform some reactions in the sequence consecutively; transformations went to completion before the next step was initiated.

Arguably the most impressive compatible cascade to date is the one-pot synthesis of a complex precursor of vitamin B_{12}, hydrogenobyrinic acid (**11**) in 12 enzymatic steps by Scott et al. (Fig. 4.3) [18]. All enzymes were prepared by overexpression of the specific biosynthetic gene of *Pseudomonas denitrificans*. After concluding that the nine enzymes worked separately in their single corresponding step, all enzymes were combined in the presence of 5-aminolevulinic acid (**10**) and the necessary cofactors. Spectacularly, complex intermediate **11** was obtained in 20% yield. After four additional chemical steps, vitamin B_{12} could be readily prepared.

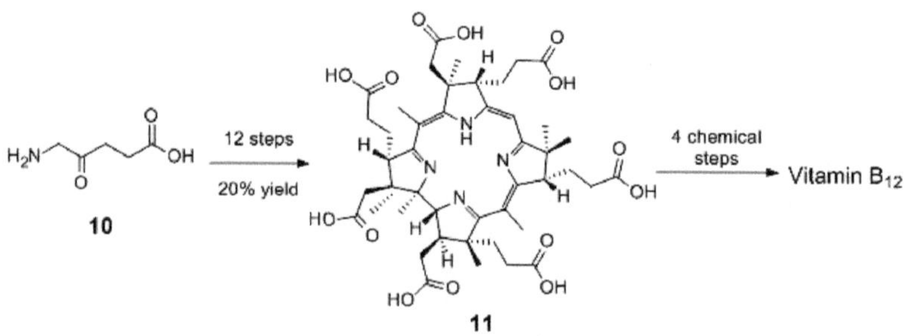

Figure 4.3 One-pot multistep synthesis of a vitamin B_{12} precursor [18].

Many reactions, however, can only be integrated if they fulfill certain requirements. It is obvious that the combined reactions should proceed in the same reaction medium and—for aqueous solutions—in a similar pH window. Generally, combining enzymes with metal complexes is problematic, as most metal complexes are not active or even decompose in water, while many enzymes are inactive in organic solvents. Furthermore, general problems with metal catalysts are poisoning of the metal center by undesired coordination of reactants or, if more than one metal catalyst is employed, ligand exchange. Another prerequisite of one-pot chemistry is that reagents or catalysts of different steps are unreactive to each other but only interact with the substrate. The combination of opposing reactions, for example, acid- and base-catalyzed or reduction and oxidation reactions, poses an additional challenge for one-pot chemistry [19]. Some successful combinations have been developed with enzymes in organic media [20], especially in the field of dynamic kinetic resolution (DKR) [12] and of water-soluble metal catalysts combined with enzymes in aqueous reaction mixtures [21, 22]. However, in general most reactions cannot be combined in a one-pot process without encountering problems due to incompatibility issues.

Several techniques have been applied to protect reagents or catalysts in one-pot reactions. Generally, like in nature, physical contact between incompatible reagents or catalysts is prevented, mostly via compartmentalization of compounds. The incompatible reagents or catalysts are confined in the compartment, while the substrate and other nonharmful reagents can move from one compartment to another. An additional advantage of compartmentalization is that catalysts can be selectively recovered from the reaction mixture in many cases. In an early stage of compartmentalization research, the chemical society focused on macroscopic separation techniques to prevent inactivation in cascade reactions. An important traditional example is compartmentalization via biphasic reaction media [23, 24], which is also widely applied in industry [25]. Other well-studied techniques are membrane reactors [26–28], sol–gels [29–31], and immobilization on supports [32–34]. Although the research about macroscopic separation methods continues, the central focus has shifted to nanoscopic compartmentalized objects in the past decade. Generally, those objects can be seen as artificial cells, as they are nanocompartments surrounded by a (porous) membrane, which are able to perform a (bio)catalytic transformation. In this chapter, we will discuss the recent developments in compartmentalization strategies for the construction of artificial cells. In our view, an important target of artificial cells is to create nanosystems that allow performing (incompatible) catalytic multistep processes. To this end, the emphasis in this chapter is on examples in which nanocompartments are used in cascade chemistry. We will also include those methods that until now have only been employed in monocatalytic processes but have the potential to be also exploited in cascade reactions.

4.2 Microdroplets

Microdroplets are aqueous spheres in oil with sizes comparable to living cells [35], which is the first important demand for the preparation of artificial cells. These small compartments are prepared by the dispersion of an aqueous solution into oil, resulting in a biphasic system (Fig. 4.4). The stability can be increased using surfactants, and droplets are ensured for prolonged periods. Enzymes, reagents, metal catalysts, or whole cells were enclosed in these droplets and still showed activity. As the aqueous layer is completely enclosed by the oil phase, encapsulation efficiencies with this technique are quantitative.

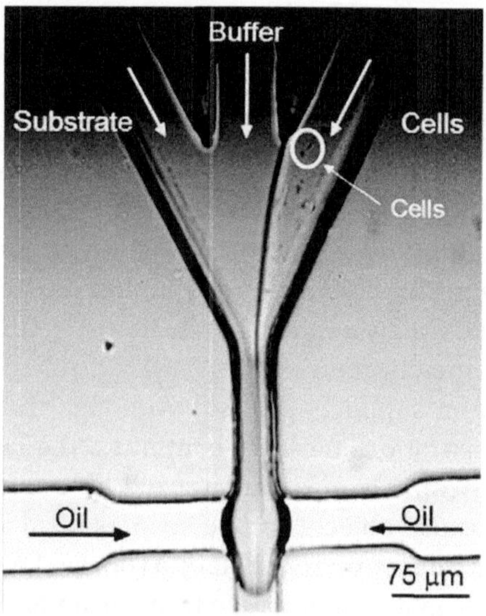

Figure 4.4 Microdroplet formation using a microfluidic device. Reproduced with permission from Ref. [36]. Copyright (2008) American Chemical Society.

The level of control over composition and also viscosity and size of the microdroplets has led to an increased interest in employing them as nanoreactors with dimensions comparable to living cells. One of the most complex multistep (bio)chemical transformations performed in microdroplets is the in vitro transcription and translation (IVTT) strategy [36, 37]. The first step, the transcription process, involves mRNA formation from encapsulated DNA. Next, the mRNA is decoded to produce the desired protein during the translation step. High-throughput measurements on protein expression and function are possible as large numbers of microscale droplets can be produced within short periods of time. Each droplet can be considered as an isolated reaction vessel, and different reaction conditions can be easily screened. Hence, microdroplets can be used in directed evolution technology, that is, the gene is randomly mutated in each droplet in order to obtain a large library of different mutated proteins that can have desirable properties [38]. Although reactions can be performed on a small scale, like in nature, microdroplets lack a permeable membrane and no diffusion occurs inside or outside the aqueous sphere. Hence, all reagents and catalyst should be present within each droplet, and incompatible processes cannot be separated using microdroplets alone.

4.3 Compartmentalization in Polymeric Nanoparticles and Micelles

Polymeric nanoparticles and micelles are small particles having a semipermeable barrier. Although this is not a membrane, compartmentalization is achieved: a sensitive reagent or catalyst is separated by enclosing it inside the polymeric bulk. In contrast to microdroplets, diffusion of substrates from the environment to the interior is possible. McQuade used polymer nanoparticles to achieve compartmentalization in a normally incompatible reaction [39, 40]. First, benzaldehyde (**12**) was reacted in *trans*-β-nitrostyrene (**13**) under (poly)ethylenimine-catalyzed conditions. Next, a nickel-mediated Michael addition produced the desired material **16** (Fig. 4.5). The combination of both catalysts in one pot without any further actions resulted in only 5% yield. This can be attributed to poisoning of the metal catalyst **14** due to coordination of the polymer nitrogen atoms with nickel. Nevertheless, by crosslinking the outer amines of the polymer, nanocapsules were formed. The capsule surface appeared impermeable for the metal catalyst, while the starting material **12** was able to penetrate and react with nitromethane to intermediate **13**.

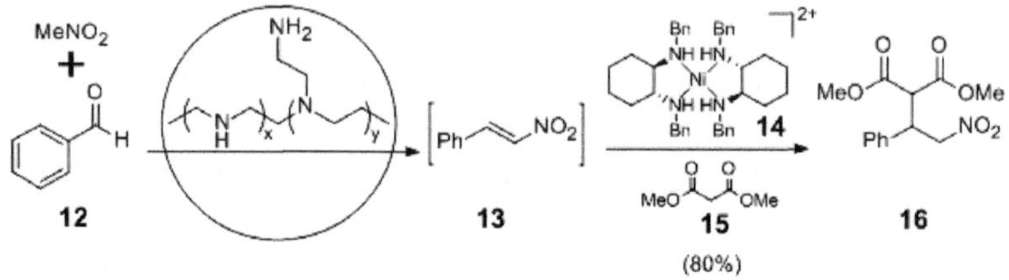

Figure 4.5 Double-catalyzed Michael adduct formation in one pot [39, 40].

Fréchet et al. used star polymers to protect acid and base catalysts from neutralization in a homogeneous mixture [41]. The acid p-toluene sulfonic acid (PTSA) and the base 4-(dimethylamino) pyridine (DMAP) were covalently attached to a styrene monomer, which was used in a crosslinking reaction with divinyl benzene and nitroxide-functional polystyrene chains to obtain star polymers. Thus, the reactive functionalities were located at the core of the materials, while the outer part contained polystyrene. It was hypothesized that the acidic and basic moieties would have no interaction as the reactive centers were unable to physically come in contact with each other (Fig. 4.6). Indeed, a yield of 65% was obtained in a cascade reaction in which an aldehyde was formed after an acid-mediated acetal hydrolysis and subsequent base-promoted Baylis Hillman reaction. If the nonadapted catalysts were used, no desired product was obtained. More importantly, also reactions in which the acid and base were attached to linear polymers yielded no desired material. This result shows that it was essential to locate the catalytic site in the interior of the star polymer.

Polymer micelles are related to the star polymers described by Fréchet [42]. Polymeric amphiphiles self-assemble above the critical micelle concentration under aqueous conditions, thereby preventing contact between water and the hydrophobic part of the polymer. Nanocompartments are thus formed with an opposite polarity as the solvent. This feature has been used to perform organic reactions in aqueous medium, which were not possible in water alone [43]. In addition, enhanced catalytic activities in polymeric assemblies have been observed, which can be attributed to high local substrate and catalyst concentrations inside the vesicles. For example, $Sc(OTf)_3$-catalyzed aldol reactions inside a micellar vesicle resulted in a fivefold increase in product formation [44]. In other research, a rhodium-catalyzed hydroformylation showed good yield in the presence of surfactants, while no conversion was observed in pure water [45]. In the last example, an enantioselective L-proline-catalyzed aldol reaction was performed inside a polymer micelle, which even showed an enhanced reaction rate compared to the reaction in an organic solvent with an unsupported catalyst [46]. Polymer micelles have also been used for recycling purposes. An encapsulated gold catalyst showed good activity in a cycloisomerization reaction and could be recycled by extraction, with a metal loss of only 0.1% or less per cycle [47]. Other examples of metal catalysis in polymeric micelles reported by the same group consist of metathesis chemistry with a Grubbs catalyst [48] and a palladium-catalyzed Sonogashira coupling [49]. Enclosing catalysts or reagents in different polymeric micelles or at different locations of the micelles would allow performing cascade reactions that are normally incompatible.

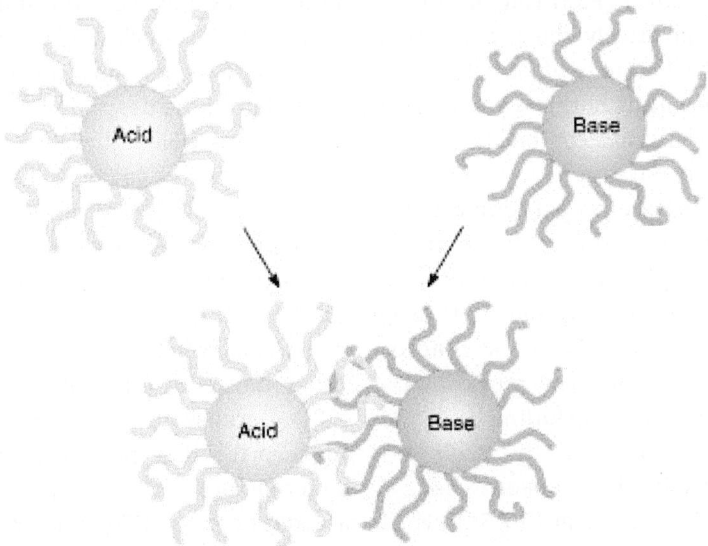

Figure 4.6 Acid- and base-containing star polymers without interaction abilities. Reproduced with permission from Ref. [41]. Copyright © 2005 WILEY-VCH Verlag GmbH & Co. KGaA, Weinheim.

4.4 Templated Nanocapsules

Particles with an internal reaction volume can be prepared by the preparation of a permeable shell around a sacrificial template. Removal of this temporary core leaves a capsule with a hollow interior, which is perfectly suited to perform (bio)catalysis. The most applied method for the creation of the shell is the layer-by-layer (LbL) technique, in which multiple thin layers are prepared step by step, that is, each layer after another. As a result of this fabrication technique, an excellent control on size, stability, and permeability is possible. Furthermore, for each layer a different material can be used as long as the different layers have a reasonable interaction to secure binding of sequential layers [50]. Polyelectrolyte capsules, which consist of consecutive multiple layers of polyanions and polycations, are an important class of LbL particles [51]. Formation of the layers is possible by stabilization by electronic interactions, which results in a highly permeable membrane [52]. In a representative example, the starting point is a colloidal particle, which is encapsulated by negatively charged polymers, poly(styrene sulfonate) (PSS) (Fig. 4.7) [53]. The next step involves addition of the positively charged poly(allyl amine) (PAH), which binds to the particle due to electrostatic interactions. The cycle of absorption of alternating charged polymers is repeated until the desired strength and thickness of the layer are achieved. Destruction of the solid core of the particle results in a hollow polyelectrolyte capsule. Problems during the preparation are mixing of the opposing charged layers, difficulties in removal of the nonbound polyelectrolytes, and incomplete coverage of the surface, all leading to vesicles with less defined and unpredictable properties.

Besides the great interest in polyelectrolyte nanocapsules for drug delivery purposes [54], other molecules have also been entrapped, including enzymes [53]. Because of the intrinsic permeability of the shell it is no surprise that polyelectrolyte capsules have been used as nanoreactors, for example, by Sukhorukov et al. [55, 56], even in cascade chemistry [57]. A normally incompatible DKR was successfully performed (Fig. 4.8). Compartmentalization of the acidic zeolite was achieved by encapsulation via the LbL method, and the nanoreactor was used together with the lipase *Candida antarctica* lipase B (CalB). The acid catalyzed the racemization of 1-phenylethanol (**17**). The (R)-enantiomer was selectively acetylated by CalB, which afforded 70% of **18** in 87% ee. Without compartmentalization, only 41% conversion and 11% ee of the acetylated alcohol **18** were observed, caused by enzyme inactivation by the acidic catalyst.

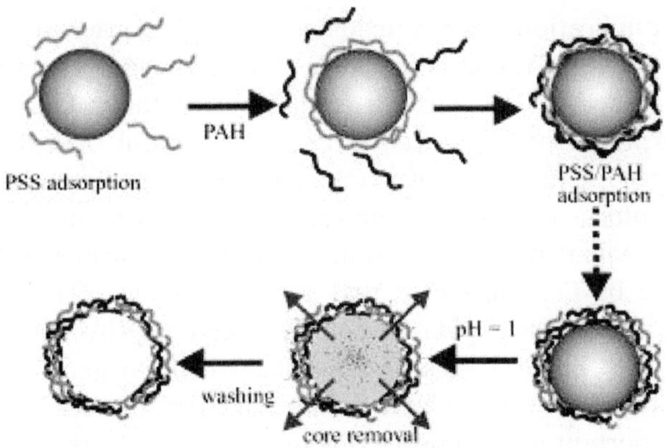

Figure 4.7 Polyelectrolyte capsule formation. Reproduced with permission from Ref. [53]. Copyright © 2004 WILEY-VCH Verlag GmbH & Co. KGaA, Weinheim.

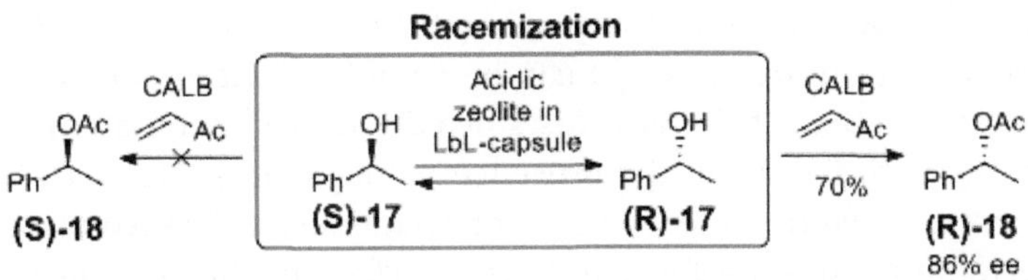

Figure 4.8 DKR with normally incompatible acidic zeolite and lipase CALB [57].

Another cascade process with LbL particles was described by the group of Li. [58]. Hemoglobin (Hb) and glucose oxidase (GOx) were used as different layers in LbL capsules. In a cascade process, glucose was oxidized by GOx with gluconic acid and H_2O_2 as products. In the next step, H_2O_2 was used by Hb for the formation of a fluorescent dye. These multienzyme loaded particles can be used as microsensors for glucose, as a higher fluorescent signal appears at elevated glucose concentrations. In addition, the researchers found that the layers showed higher permeability at higher glucose concentrations. Hence, these particles are good candidates for advanced drug delivery devices because the cargo can be delivered at locations with high glucose concentrations. In later work, Li and coworkers prepared vesicles with GOx and catalase shells loaded with insulin [59]. Upon glucose addition, the permeability was increased, which led to insulin release.

The duplication of DNA inside a capsule is a great step toward an artificial cell. Trau et al. performed a polymerase chain reaction (PCR) inside an LbL vesicle [60]. Thermostable LbL particles were formed, which is a requisite for successful DNA formation. Polymerase, templates, and primers were encapsulated, while the building blocks of the DNA, the nucleotides, were able to pass the semipermeable membrane. Another important achievement was described by Caruso, which was the production of RNA via the transcription of encapsulated double-stranded DNA inside an LbL vesicle [61]. This process is an essential part of the gene expression in cells, and this work is a further step toward a mimic of a living cell. Other important natural processes that were conducted within LbL capsules are DNA degradation [62] and ATP formation [63, 64].

A higher level of complexity in compartmentalization was achieved by the group of Möhwald with the construction of vesicle-in-vesicle architecture [65]. The so-called shell-in-shell microcapsules are multicompartment nanoreactors in which two normally incompatible steps can be combined *within one particle*. The preparation of the vesicles started with coprecipitation of an enzyme with $CaCO_3$ (in the presence of magnetite) (Fig. 4.9). Next, the particles were coated via the LbL technique, and a second $CaCO_3$ precipitation was performed at the surface in the presence of a second enzyme. At this time, no magnetite was added which enables the purification of the desired material from the coproduct, which consisted of $CaCO_3$ particles that were formed spontaneously in solution and did not encapsulate the previously made LbL particles. Another coating was installed, followed by removal of both inorganic supports. The compartments of the microvessels were independently loaded with GOx and peroxidase. In a cascade reaction, enzymatic oxidation of glucose yielded H_2O_2, which was used by peroxidase to produce a highly fluorescent species. A fluorescent signal was only observed in the peroxidase compartment of the vesicle. This result shows that no leakage of the enzyme occurred, and peroxide was able to diffuse from the glucose oxidation compartment to the other compartment. This concept was expanded by Bäumler, who encapsulated three enzymes in three different compartments of one single, shell-in-shell-in-shell, particle [66].

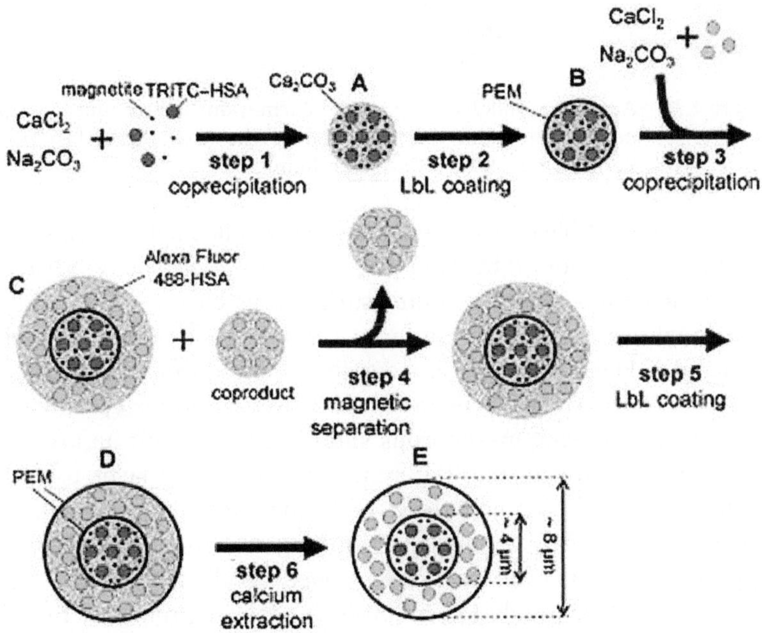

Figure 4.9 Formation of shell-in-shell microcapsules by Möhwald et al. [65]. Reproduced with permission from Ref. [65]. Copyright © 2007 WILEY-VCH Verlag GmbH & Co. KGaA, Weinheim.

Related materials are yolk-shell vesicles. These are vesicles with a shell and a heterogeneous nanoparticle in the core, which, however, does not utilize the whole inner compartment. An early example of a catalytic application has been described by the group of Song [67]. This material was prepared using a simple approach, which started with the preparation of gold nanoparticles, which were coated by silica. In the last step, the metal was partially etched away in a controlled manner, leaving an empty space between the metallic core and the silica shell (Fig. 4.10). The particles showed catalytic activity in the reduction of *p*-nitrophenol by $NaBH_4$. The heterogeneous catalyst stayed inside the compartment, while the substrate and the reducing agent were able to diffuse through the silica shell.

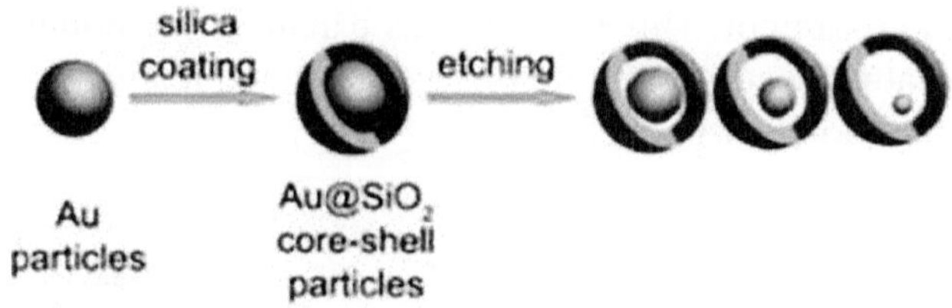

Figure 4.10 Au@SiO$_2$ yolk/shell particle formation. Reproduced with permission from Ref. [67]. Copyright © 2008 WILEY-VCH Verlag GmbH & Co. KGaA, Weinheim.

An incompatible cascade sequence using yolk-shell particles has been recently reported by Yang (Fig. 4.11) [68]. The core consisted of a basic amine, while the outer surface was covered with acidic functionalities. The preparation was a combination of the methods depicted in Figs. 4.9 and 4.10: A sacrificial silica layer was attached to a solid nanoparticle functionalized with boc-protected amines. Next, a thiol-bearing mesoporous silica shell was built up around this silica-coated system. After removal of the silica layer, a yolk-shell particle was obtained. Next, the amines were deprotected and the thiol groups oxidized to SO_3H, resulting in particles bearing two opposing catalysts. The particles showed excellent activity in an acid-catalyzed deacetylation followed by a base-promoted Henry condensation. Additionally, the material could be recycled with only a slight loss in activity.

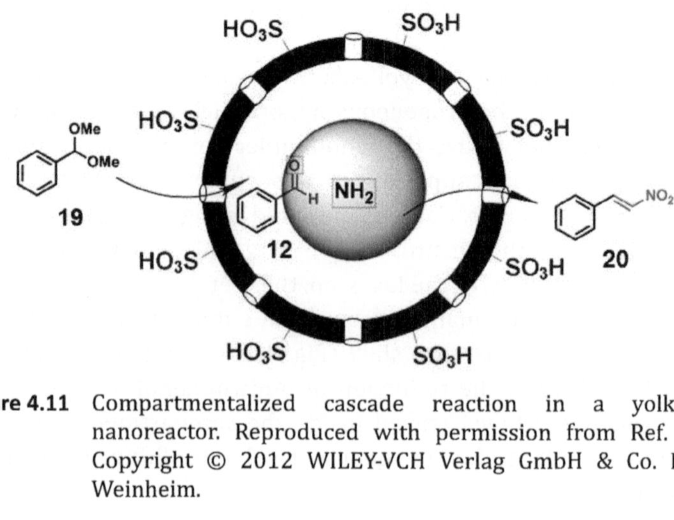

Figure 4.11 Compartmentalized cascade reaction in a yolk-shell nanoreactor. Reproduced with permission from Ref. [68]. Copyright © 2012 WILEY-VCH Verlag GmbH & Co. KGaA, Weinheim.

4.5 Liposomes

Liposomes are vesicles with a permeable bilayer of phospholipids. Due to the amphiphilic nature of the phospholipids, nanocompartments with dimensions between 20 and 200 nm are formed in water to minimize hydrophobic–hydrophilic interactions. Liposomes have been well studied in the past, and current interest focuses on mimicking the living cell [69], enzyme encapsulation and stabilization [70], and drug delivery [71].

The concept of compartmentalization is nicely illustrated in an example described by Moore [72]. ATP was produced in a liposome after incorporation of both ATP synthase and a proton pump in the liposomal membrane of the vesicle (Fig. 4.12). ATP synthase is a membrane protein with two subunits, which is able to produce ATP from ADP and inorganic phosphate. Electrons can be transferred through the membrane via the carotene–porphyrin–naphthoquinone amphiphile after excitation (C-P-Q to $C^{\bullet+}$-P-$Q^{\bullet-}$) and a lipophilic quinone (Qs) in an alternating oxidized–reduced form (Fig. 4.12). Upon a pH difference between the two sides of the membrane, one of the subunits of the synthase starts to rotate and produces ATP. One of the subunits of ATP synthase was embedded inside the lipid membrane of a liposome, while the other subunit could freely rotate and hence produce ATP.

The production of proteins via the IVTT strategy inside these lipid vesicles is another impressive example of a cascade process and an alternative for protein formation from cell extracts [73]. The different reaction stages have been already explained in the microdroplet section. In the group of Yomo, green fluorescent protein (GFP) was produced in such a cascade process [74]. First, a gene was expressed to obtain RNA polymerase, which transcribed a second gene, which yielded the right-folded GFP. In a second impressive example, Libchaber et al. expressed α-hemolysin, a pore-forming protein, inside liposomes via the transcription–translation sequence. The in situ prepared protein migrated to the shell and formed a membrane channel for nutrients for another protein formation process [75].

Recently, the group of Zasadzinski reported a multicompartment liposome [76]. Large vesicles with a lipid bilayer were formed in the presence of smaller ones, which yielded liposomes in liposomes or so-called "vesosomes." Encapsulation of a drug inside the vesosomes resulted in an enhanced stability in serum. The half-life of ciprofloxacin, a drug often used in drug release studies, increased from 10 minutes in single liposomes to 6 hours in vesosomes. This can be attributed to the double lipid membrane that has to be crossed instead of a single membrane in liposomes. This multicompartment vesicle is the first step toward more complex systems and a closer mimic of a natural cell. The multiple organelles in living cells can be considered as inner compartments that are necessary to achieve an efficient metabolism. Compartmentalization within one particle will have benefits compared to multiple single-compartment vesicles in chemical transformations, as intermediates need to travel a shorter distance to the next reactive site.

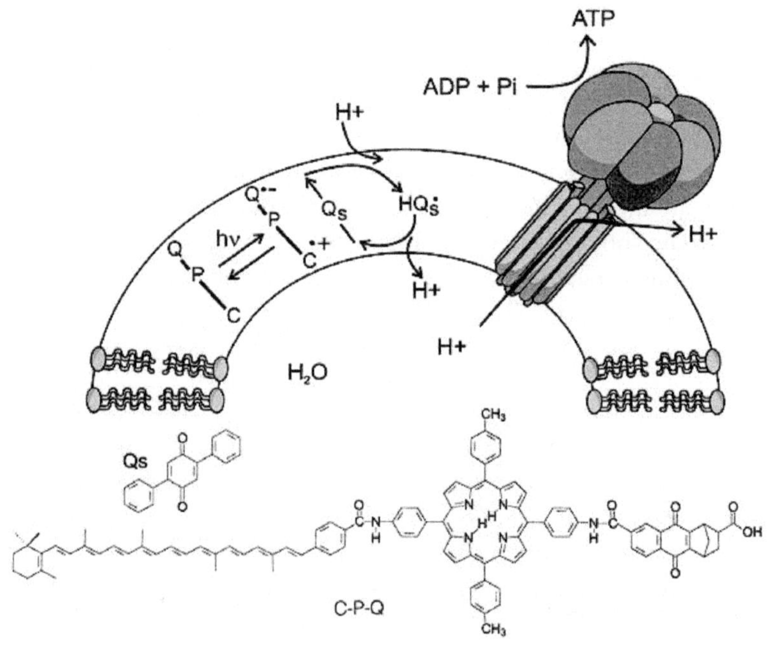

Figure 4.12 ATP formation at the surface of a liposome ($C^{\bullet+}$-P-$Q^{\bullet-}$ represents the excited amphiphile). Reproduced from Ref. [72] with permission from Macmillan Publishers Ltd., copyright (1998).

Caruso and coworkers designed liposome-in-polymersome vesicles, called capsosomes, for nanocarrier purposes [77]. First, a cholesterol-based polymeric film was formed around silica nanoparticles, which gave a stable absorption of phospholipid liposomes. A next layer was applied by adding a mixture of a cholesterol-based polymer and a different, crosslinkable polymer around the liposomes. Stable capsules were obtained after crosslinking of the outer polymer shell. Although this system was designed for transport of the liposome loading, it should also be applicable for cascade catalysis in which each catalyzed transformation is performed in a different compartment.

4.6 Polymersomes

Polymersomes are the synthetic analogues of liposomes and were firstly reported by Discher in 1999 [78]. Inspired by the self-assembly behavior of natural phospholipids in water, synthetic polymeric amphiphiles were added to water and robust vesicles were spontaneously formed. Dyes and proteins were successfully encapsulated in those nanocompartments. Polymersomes are generally more stable with respect to liposomes, which is attributed to the higher molecular weight of the polymers [79]. In addition, properties of polymersome are better tunable, as many different hydrophobic and hydrophilic polymers can be used and the polymers can be synthetically adapted [80]. Also the size of the hydrophobic and hydrophilic parts and the length of the whole polymer have an influence on the thickness, strength, and porosity of the vesicle membrane.

Besides proposed biomedical applications of polymersomes as drug carriers [81], these vesicles have also been extensively used in compartmentalized catalysis. The polymeric membrane can be used as an efficient semi-permeable gate: small molecules can easily enter the compartment, while large materials like enzymes or heterogeneous nanoparticles stay inside. Hence, polymersomes would be useful to perform cascade reactions with incompatible catalysts, provided that these catalysts are unable to cross the membrane. To date, no incompatible reactions have been combined with the aid of polymeric nanoreactors as the compartmentalization technique. However, examples of compatible reactions that were combined in a one-pot reaction will be discussed to show the concept of compartmentalization via polymersomes.

Most polymersome membranes are impermeable for even small molecules [82], which limits the amount of substrate available for a catalytic reaction to the initial quantity of enclosed starting material [83, 84]. Nevertheless, these nonpermeable capsules were used to investigate the behavior of enzymes in small volumes, which showed an enhancement of enzymatic activity. This was attributed to the finite volume of the reactions, which increased the number of collisions between the reagents.

Polymersomes prepared from poly(2-methyloxazoline)-block–poly(dimethylsiloxane)-block–poly(2-methyloxazoline) (PMOXA-PDMS-PMOXA) are also only permeable to rather small molecules as superoxide (O_2^-) and O_2 [85]. The first step to enhance the permeability was the introduction of ionophores in the membranes [86]. These lipophilic ion carriers were able to transport Ca^{2+} over the membranes, after which it crystallized with PO_4^{2-} inside the vesicles. Organic substrate-permeable membranes in (PMOXA-PDMS-PMOXA) polymersomes were achieved by incorporating a bacterial channel protein, OmpF, in the membrane [87, 88]. Metabolites up to 400 Da are able to pass those channels in nature. The successful creation of permeable membranes was demonstrated by applying ampicillin (MW 350) to β-lactamase loaded polymersomes, which should hydrolyze ampicillin to ampicillinoic acid. Indeed, the formation of the acid was observed, while with polymersomes that lack the OmpF channel, no conversion was observed. In the research that followed, also another protein channel was installed, Tsx, which selectively transfers nucleosides and nucleotides through membranes. After encapsulation of the nucleoside hydrolase of *Trypanosoma vivax* (TvNH), the three natural substrates (inosine, adenosine, and guanosine) were tested. Indeed, the starting materials were able to diffuse through the membrane and were converted to the intended products. Additionally, three prodrugs were tested, which were readily converted to cytotoxic compounds. This research showed that it is possible to use these vesicles as drug delivery systems: an active drug can be in vivo prepared at a specific target location, thereby preventing decomposition of the labile drug in an early stage after ingestion [89]. This type of polymersome was further improved by the incorporation of a more efficient protein channel, which allowed a faster movement across the membrane [90].

In other work, permeable polymersomes were created with mixtures of amphiphilic polymers [91]. After polymersome formation, one of the two types of polymers was removed from the bilayer in a stimulus-responsive manner, resulting in pores in the polymer membrane. This should allow for small substrates to enter the vesicle. Indeed, when pores were introduced in CalB-loaded vesicles, enzymatic activity was observed. On the other hand, nontreated polymersomes did not show biocatalytic activity, demonstrating that the permeability was introduced by the selective removal of one of the polymers. Recently, Battaglia, Voit, et al. reported polymersomes containing membranes with pH-switchable permeability [92]. Polymersomes were prepared from poly(ethylene glycol)-diethyl amino ethylmethacrylate (PEG-DEAEM) polymers. The diethylamino moiety in the hydrophobic part could be protonated, which resulted in a more hydrophilic inner part of the membrane. In order to maintain polymersome integrity, the membrane structure was crosslinked before protonation. Hence, the membranes became permeable to small molecules, and enzymatic reactions could be carried out within the vesicles. Permeability could also be induced by applying an external electrical field, called electroporation [93]. This is a well-known technique for the introduction of macromolecules to living cells. It was shown that this method was also applicable for polymersomes. The electrical field temporarily destabilized the polymersome membrane, which allowed (macro)molecules to move through the membrane.

In contrast to the previous examples, the polymersomes described by Li and coworkers have intrinsically permeable membranes [94]. In this work, L-asparaginase was encapsulated with 80% efficiency. An enhanced stability and a small increase in enzymatic activity were observed, which was a result of a stable environment created by the polymersomes. Without any further actions to introduce membrane permeability, the enzyme successfully converted L-asparagine to L-aspartic acid inside the polymeric nanoreactors.

Much research has been described with polymersomes prepared from the diblock copolymer polystyrene$_{40}$-b-poly(L-isocyanoalanine(2-thiophen-3-yl-ethyl)amide)$_{50}$ (PS-PIAT) [95]. This membrane is intrinsically permeable to small molecules. Thus, no additional actions are necessary to induce permeability in order to use them as nanoreactors. Successful ring-opening polymerizations of lactones with polymersomes loaded with CalB in either the lumen or the hydrophobic bilayer were performed [96]. It was shown that if the enzyme was located in the membrane instead of the lumen, shorter oligomers were formed. This phenomenon was explained by the presence of steric bulk in the membrane.

The possibility of performing cascade chemistry has been demonstrated with GOx and horseradish peroxidase (HRP) [97]. First, both enzymes were each enclosed in separate polymersomes. After mixing the polymersomes in a 1:1 ratio, a cascade reaction was performed in which first glucose was oxidized, which liberated H_2O_2. Then, the reactive oxidant passed the membrane of the HRP-loaded nanocapsules, which resulted in enzymatic oxidation of 2,2'-azino-bis(3-ethylbenzothiazoline-6-sulphonic acid) (ABTS) (Fig. 4.13). The successful compartmentalization of both enzymes was demonstrated by the addition of subtilisin A, which is a proteolytic enzyme that degrades enzymes if physical contact is possible. After the addition of the proteolytic enzyme to the outer volume of the reaction mixture, still oxidation of ABTS was observed. This result showed that both enzymes remained active and that the polymersome membrane acted as a protected nanocompartment.

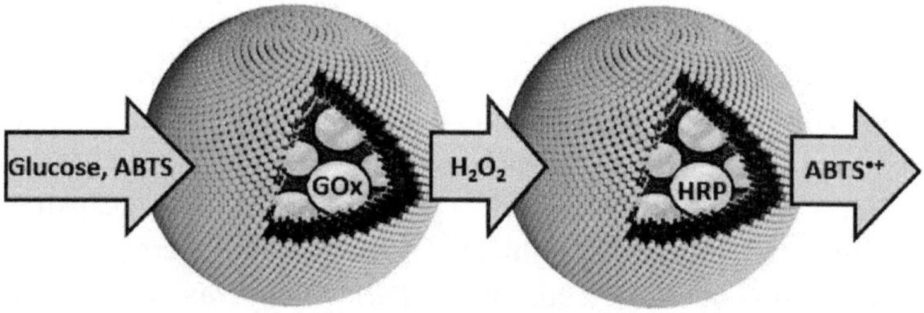

Figure 4.13 Cascade reaction with two different-loaded polymersomes [97].

In the next step, three enzymes were combined in a one-pot process. Two enzymes were installed inside a single polymersome, one in the lumen and one in the membrane. The third enzyme was first kept outside the polymersome and in later work attached to the membrane surface (Fig. 4.14) [98, 99]. The first enzymatic step comprised a CalB-catalyzed deacetylation of glucose acetate in the membrane of the nanoreactor. Next, the deprotected alcohol was selectively oxidized by GOx, which produced hydrogen peroxide. This compound diffused through the membrane and was used in the oxidation of ABTS by the surface-attached HRP, yielding a fluorescent species. This combination of enzymes is also active in solution, but it nicely shows the possibility of compartmentalization. Hence, the same strategy would be applicable for incompatible enzymes or metal catalysts as well.

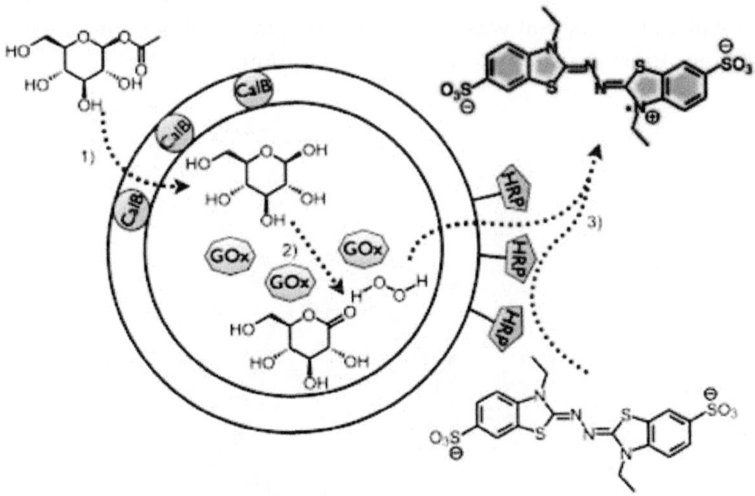

Figure 4.14 Three-enzyme-mediated cascade reaction with all enzymes attached to one PS-PIAT polymersome. Reproduced with permission from Ref. [99]. Copyright © 2009 WILEY-VCH Verlag GmbH & Co. KGaA, Weinheim.

The possibility to perform compartmentalized cascade catalysis involving cofactor regeneration was also demonstrated for the PS-PIAT polymersomes. Glucose-6-phosphate dehydrogenase (G6PDH) was enclosed in a polymersome [100]. This enzyme is capable of converting NADPH into $NADP^+$, a reaction of high importance in the field of biocatalysis, as many enzymes require $NADP^+$ as a cofactor. The formation of $NADP^+$ was combined with phenylacetone monooxygenase (PAMO) in a cascade process, resulting in an efficient Baeyer–Villiger reaction.

In an early and impressive example of enhanced complexity, Montemagno prepared a multienzyme-bearing polymersome (Fig.

4.15) [101], analogously to the liposome example of Moore [72], and discussed in the corresponding section. Under the influence of light, protons were continuously introduced in the polymersome via the enzyme bacteriorhodopsin (BR), which was also installed in the polymeric membrane. A pH difference of almost 0.1 was observed between the interior and exterior of the vesicle, which was large enough to produce ATP. In the absence of light, no ATP production was detected, showing that it was essential to have two active proteins. This design was impossible to conduct without a vesicle. However, its function was to create a proton-enriched environment and not to protect enzymes from deactivation.

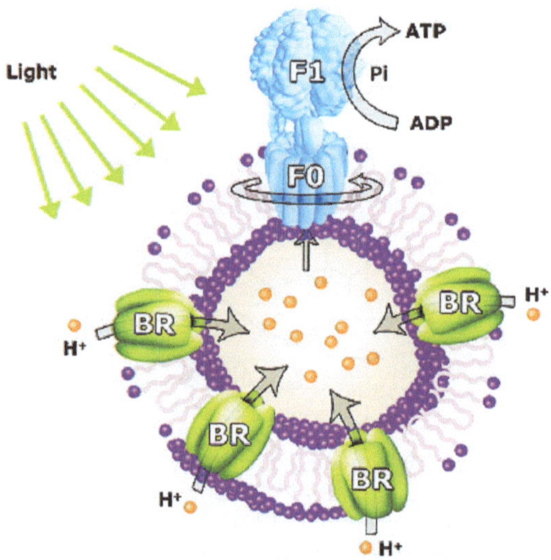

Figure 4.15 BR–ATP synthase–mediated ATP formation based on a proton gradient. Reproduced with permission from Ref. [101]. Copyright (2005) American Chemical Society.

To successfully imitate the natural metabolism process in polymeric vesicles, a demand exists for multicompartmentalized vesicles [102]. Nallani prepared polymersomes-in-polymersomes by the formation of the large, outer polymeric capsules in the presence of the smaller ones [103]. Besides single polymersomes of both the large and the small type, polymersomes-in-polymersomes were also observed. This example of a vesicle with more than one compartment was supported by an experiment with a smart choice of macromolecular dyes: a red and a green fluorescent dye were encapsulated in, respectively, the large and small polymersomes. Fluorescence microscopy showed either red or green emission for the single polymersomes, while the polymersome-in-polymersome system showed a yellow emission, indicating that the two dyes are in close proximity of each other and that the small polymersomes were indeed encapsulated into the larger ones. In a more recent example, Lecommandoux also prepared polymersomes-in-polymersomes [104]. The encapsulation of the smaller vesicles into the larger ones was achieved via an emulsion-centrifugation procedure, leading to a highly efficient method to prepare multicompartmentalized vesicles.

Multicompartmentalized polymersome gels were prepared by Weitz and coworkers via a facile method [105]. First, amphiphilic polymers were added to a mixture of a volatile good solvent and a less volatile bad solvent, that is, a solvent in which the polymer has poor solubility (Fig. 4.16). Water was added and the amphiphiles adsorbed at the W/O interface to reduce unfavorable hydrophobic–hydrophilic interactions. Next, the good solvent was slowly evaporated, giving adhesion of the polymersomes and formation of a multicompartmentalized polymersome gel. Different dyes could be enclosed in this gel without cross-contamination. In this case, the polymer mixtures were separately mixed with the different dyes and then added to the same aqueous medium. This system is suitable for compartmentalized catalysis, as the catalysts of the separate steps can be enclosed in different compartments. However, no catalysis with these systems has been reported so far.

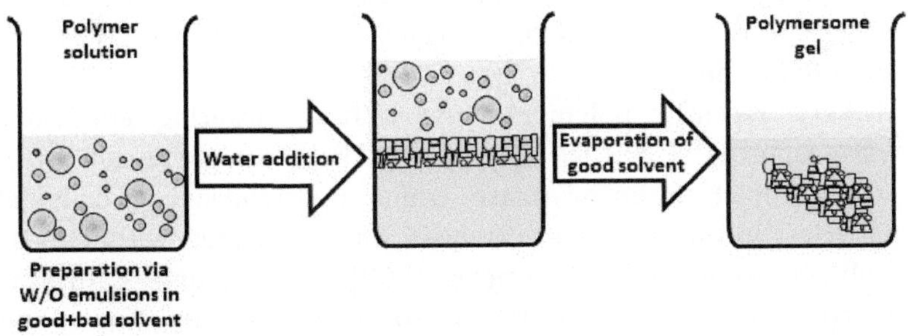

Figure 4.16 Polymersome–gel formation via a straightforward W/O emulsion method [105].

The same group prepared vesicles which can be seen as two spheres adhering together [106]. Formation of the multicompartment vesicles was performed in a well-controlled fashion (Fig. 4.17). First, W/O/W double emulsions with multiple inner water droplets were prepared with glass capillary microfluidics. A di-block copolymer stabilized the interfaces between the organic solvent and water. Upon evaporation of the oil phase, the water droplets that were initially in the oil droplets stuck together. Importantly, the polymeric walls between the aqueous compartments remained intact, which yielded polymeric capsules with multiple compartments. These complex materials would be perfectly suited for cascade chemistry, as subsequent incompatible processes can be performed in close proximity of each other.

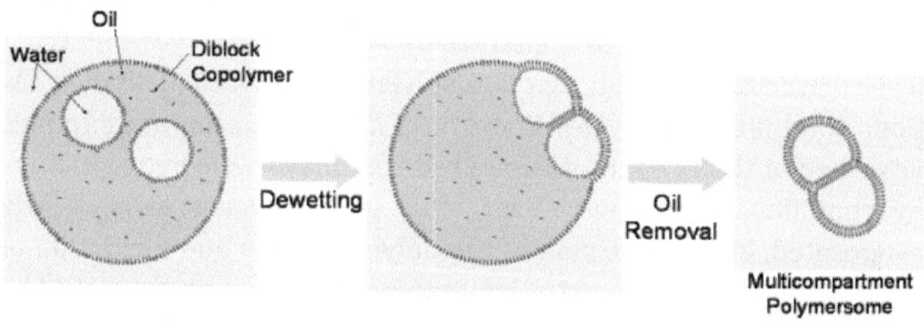

Figure 4.17 Multicompartment formation via the W/O/W double emulsion. Reproduced with permission from Ref. [106]. Copyright © 2011 WILEY-VCH Verlag GmbH & Co. KGaA, Weinheim.

In another work of the same researchers, polymersomes were prepared with a triple polymeric membrane [107]. Each membrane consisted of the same amphiphile. The multicompartmentalized vesicles were constructed following a step-by-step approach using a microfluidic device.

Very recently, polymersome-stabilized pickering emulsions were reported (Fig. 4.18) [108, 109]. These emulsions consist of water droplets in an oil phase, which are stabilized by colloidal particles. In one example, catalysis was performed within these multicompartmentalized vesicles [109]. To this end, CalB was encapsulated in either the polymersomes or the water droplet and showed enhanced yields in an unfavorable esterification reaction in water. The improved reactivity can be explained by the presence of a large hydrophobic medium and small aqueous compartments in which the lipase was active. The lipid nature of the product ensured that the molecule stayed only shortly in the aqueous vesicle.

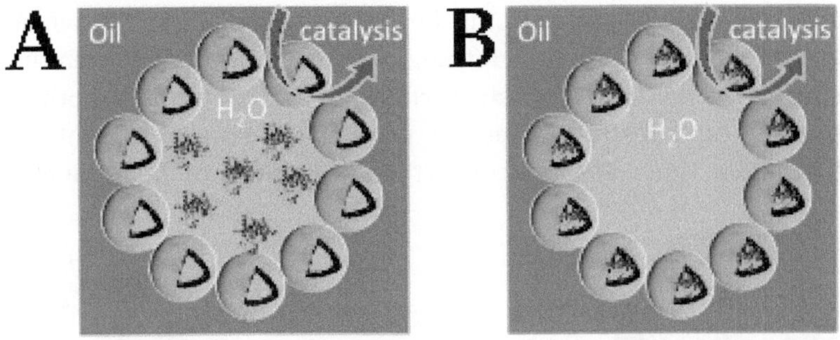

Figure 4.18 Polymersome-stabilized pickering emulsions with CALB encapsulated in the water droplet (A) or in polymersomes (B). Reproduced with permission from Ref. [109]. Copyright © 2012 WILEY-VCH Verlag GmbH & Co. KGaA, Weinheim.

4.7 Viruses and Virus-Like Particles

Virus-like particles (VLPs) are one of the best-defined capsules that have been employed as nanoreactors. Generally, these viral cages consist of multiple copies of only one or a few different protein subunits, which are assembled in a highly ordered fashion. Naturally, virus cages contain genetic material that is brought inside living host cells for reproduction of the virus. Release of this material occurs by disassembly of the protein cage, followed by reforming the virus particle. The interest in using viral nanoreactors is highly increased in the past years because these compartments can be considered as artificial organelles. The contents of the capsules are compartmentalized and hence protected from the environment. Enzyme-loaded capsules are, besides application in the field of protein carriers [110], perfectly suited for catalysis as the shell of capsules is highly permeable to substrates [111–113]. Nolte et al. have demonstrated that the natural genetic material of the capsid of the cowpea chlorotic mottle virus (CCMV) can be removed by using a reversible disassembly–assembly strategy, which depends on pH (Fig. 4.19). Since the genetic information is missing, the protein nanocontainer is referred to as a VLP. If the reassembly is performed in the presence of *Pseudozyma antarctica* lipase B (PalB), a precise number of enzymes could be encapsulated, which was shown by labeling the protein with a dye [113]. The enzyme still showed activity after the enclosure in the VLP, and the catalytic efficiency appeared to decrease with higher numbers of proteins present in a single capsule. The scientists attribute this effect to the hypothesis that only one substrate molecule is present in the capsid, and only one biocatalyst is necessary to convert the substrate into the product.

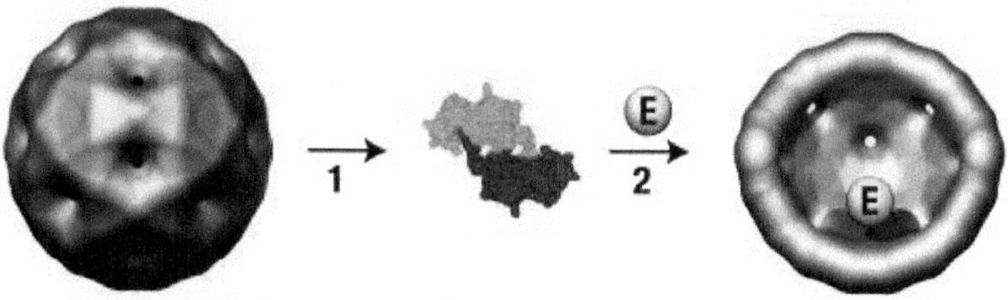

Figure 4.19 Encapsulation of guest protein (E) in a VLP. Reproduced from Ref. [111] with permission from Macmillan Publishers Ltd., copyright (2007).

In another example of a catalytic reaction within a VLP, the enzyme yeast cytosine deaminase (yCD) was fused to one of the proteins that form the shell of an SV40 capsid [114]. After formation of the VLPs, the enzyme was encapsulated in the nanocontainer and catalytically converted 5-fluorocytosine to 5-fluorouracil, the latter inducing cell death. Indeed, after addition of these adapted viruses to living cells, cell death was observed, showing that these VLPs were still able to attach the cell surface and deliver their content.

The first example of a metallic nanoparticle enclosure in a VLP was the formation of $H_2W_{12}O_{42}^{10-}$ inside the CCMV [115]. Numerous other examples appeared in the literature, but no catalysis has been described so far with these materials [116].

4.8 Outlook

The chemical community started to work on cascade reactions because potentially much higher efficiencies are obtained. Time, cost, and waste can be reduced by performing reactions in a subsequent manner. The search for compatible catalysts and reagents to perform multiple steps in one pot will continue as this approach is cheap and relatively easy to apply. However, some reactions can be impossibly combined, and compartmentalization is the key technique to overcome incompatibilities. The approach to separate harmful catalysts and reagents has been changed from a macroscopic to a nanoscopic level since the start of cascade chemistry. In early examples, separation was achieved by membrane reactors, via biphasic reaction conditions, and on heterogeneous supports. Compartmentalization at nanosized dimensions can be seen as efforts to mimic living cells. Polyelectrolyte capsules, shell-in-shell vesicles, liposomes, and polymersomes are recent examples in the field of compartmentalization and have an inner compartment surrounded with a shell or membrane. Furthermore, (bio)catalysts can be encapsulated, which allows conducting of chemical transformations within artificial cells. Due to the small dimensions of the compartments, the path from one catalytic site to another is short and subsequent reactions can proceed faster.

A better imitation of a living cell will enhance the efficiency of cascade reactions. Consequently, it is expected that the complexity of the nanoreactors will increase in the future. More steps can be incorporated in a cascade sequence if multicompartment vesicles are used, which has been already shown with shell-in-shell capsules, liposomes, and polymersomes. Additionally, catalysts can be attached at different locations in or around the vesicle. Another way to obtain a closer mimic of a cell is to incorporate selection channels in the membrane. In this way, desired substrates can selectively enter the compartment, while unreactive or harmful compounds will not occupy the compartment and drive away the substrate. The next, upcoming step would be to navigate substrates to the catalytic site. In previous examples about cascade reactions in microcapsules, substrates diffused in random directions and reacted more or less "incidentally" near the biocatalyst. Navigation of the substrates, for example, by mimicking metabolic channeling, would definitely enhance the efficiency, as the sequence of transformations will be faster.

References

1. Hurtley, S. (2009) Location, location, location, *Science*, **326**, 1205–1205.

2. Chen, A. H., and Silver, P. A. (2012) Designing biological compartmentalization, *Trends Cell Biol.*, **22**, 662–670.
3. Fernie, A. R., Carrari, F., and Sweetlove, L. J. (2004) Respiratory metabolism: glycolysis, the TCA cycle and mitochondrial electron transport, *Curr. Opin. Plant Biol.*, **7**, 254–261.
4. Raines, C. A. (2003) The Calvin cycle revisited, *Photosynth. Res.*, **75**, 1–10.
5. Randle, P. J. (1998) Regulatory interactions between lipids and carbohydrates: the glucose fatty acid cycle after 35 years, *Diabetes/Metab. Rev.*, **14**, 263–283.
6. Smirnoff, N., Conklin P. L., and Loewus, F. A. (2001) Biosynthesis of ascorbic acid in plants: a renaissance, *Annu. Rev. Plant Physiol. Plant Mol. Biol.*, **52**, 437–467.
7. Croteau, R., Ketchum, R. E. B., Long, R. M., Kaspera, R., and Wildung, M. R. (2006) Taxol biosynthesis and molecular genetics, *Phytochem. Rev.*, **5**, 75–97.
8. Winkel, B. S. J. (2004) Metabolic channeling in plants, *Annu. Rev. Plant Biol.*, **55**, 85–107.
9. Visser, W. F., van Roermund, C. W. T., IJlst, L., Waterham, H. R., and Wanders, R. J. A. (2007) Metabolite transport across the peroxisomal membrane, *Biochem. J.*, **401**, 365–375.
10. Lye, G. J., and Woodley, J. M. (1999) Application of in situ product-removal techniques to biocatalytic processes, *Trends Biotechnol.*, **17**, 395–402.
11. Bruggink, A., Schoevaart, R., and Kieboom, T. (2003) Concepts of nature in organic synthesis: cascade catalysis and multistep conversions in concert, *Org. Process Res. Dev.*, **7**, 622–640.
12. Pàmies, O., and Bäckvall, J. E. (2003) Combination of enzymes and metal catalysts. A powerful approach in asymmetric catalysis, *Chem. Rev.*, **103**, 3247–3261.
13. Zhou, J. (2010) Recent advances in multicatalyst promoted asymmetric tandem reactions, *Chem. Asian J.*, **5**, 422–434.
14. Ricca, E., Brucher, B., and Schrittwieser, J. H. (2011) Multi-enzymatic cascade reactions: overview and perspectives, *Adv. Synth. Catal.*, **353**, 2239–2262.
15. Sheldon, R. A. (2012) Fundamentals of green chemistry: efficiency in reaction design, *Chem. Soc. Rev.*, **41**, 1437–1451.
16. Noyori, R. (2009) Synthesizing our future, *Nat. Chem.*, **1**, 5–6.

17. Elders, N., van der Born, D., Hendrickx, L. J. D., Timmer, B. J. J., Krause, A., Janssen, E., de Kanter, F. J. J., Ruijter, E., and Orru, R. V. A. (2009) The efficient one-pot reaction of up to eight components by the union of multicomponent reactions, *Angew. Chem., Int. Ed.*, **48**, 5856–5859.

18. Scott, A. I. (1994) Towards a total, genetically-engineered synthesis of vitamin-B12, *Synlett*, 871–883.

19. Schrittwieser, J. H., Sattler, J., Resch, V., Mutti, F. G., and Kroutil, W. (2011) Recent biocatalytic oxidation-reduction cascades, *Curr. Opin. Chem. Biol.*, **15**, 249–256.

20. Caiazzo, A., Garcia, P. M. L., Wever, R., van Hest, J. C. M., Rowan, A. E., and Reek, J. N. H. (2009) Synergy between chemo- and bio-catalysts in multi-step transformations, *Org. Biomol. Chem.*, **7**, 2926–2932.

21. Burda, E., Hummel, W., and Gröger, H. (2008) Modular chemoenzymatic one-pot syntheses in aqueous media: combination of a palladium-catalyzed cross-coupling with an asymmetric biotransformation, *Angew. Chem., Int. Ed.*, **47**, 9551–9554.

22. Mutti, F. G., Orthaber, A., Schrittwieser, J. H., de Vries, J. G., Pietschnig, R., and Kroutil, W. (2010) Simultaneous iridium catalysed oxidation and enzymatic reduction employing orthogonal reagents, *Chem. Commun.*, **46**, 8046–8048.

23. Horváth, I. T., and Rábai, J. (1994) Facile catalyst separation without water: fluorous biphase hydroformylation of olefins, *Science*, **266**, 72–75.

24. Grigg, R., and York, M. (2000) Bimetallic catalytic cascade ring closing metathesis: intramolecular Heck reactions using a fluorous biphasic solvent system or a polymer-supported palladium catalyst, *Tetrahedron Lett*, **41**, 7255–7258.

25. Cornils, B. (1998) Industrial aqueous biphasic catalysis: status and directions, *Org. Process Res. Dev.*, **2**, 121–127.

26. Wichmann, R., Wandrey, C., Bückmann, A. F., and Kula, M. R. (1981) Continuous enzymatic transformation in an enzyme membrane reactor with simultaneous Nad(H) regeneration, *Biotechnol. Bioeng.*, **23**, 2789–2802.

27. Dijkstra, H. P., van Klink, G. P. M., and van Koten, G. (2002) The use of ultra- and nanofiltration techniques in homogeneous catalyst recycling, *Acc. Chem. Res.*, **35**, 798–810.

28. Janssen, M., Müller, C., and Vogt, D. (2011) Recent advances in the recycling of homogeneous catalysts using membrane separation, *Green Chem.*, **13**, 2247–2257.

29. Hench, L. L., and West, J. K. (1990) The sol-gel process, *Chem. Rev.*, **90**, 33–72.
30. Lu, Z. L., Lindner, E., and Mayer, H. A. (2002) Applications of sol-gel-processed interphase catalysts, *Chem. Rev.*, **102**, 3543–3577.
31. Gelman, F., Blum, J., and Avnir, D. (2001) Acids and bases in one pot while avoiding their mutual destruction, *Angew. Chem., Int. Ed.*, **40**, 3647–3649.
32. Cohen, B. J., Kraus, M. A., and Patchornik, A. (1981) Wolf and lamb reactions: equilibrium and kinetic effects in multipolymer systems, *J. Am. Chem. Soc.*, **103**, 7620–7629.
33. Phan, N. T. S., Gill, C. S., Nguyen, J. V., Zhang, Z. J., and Jones, C. W. (2006) Expanding the utility of one-pot multistep reaction networks through compartmentation and recovery of the catalyst, *Angew. Chem., Int. Ed.*, **45**, 2209–2212.
34. Shylesh, S., Wagener, A., Seifert, A., Ernst, S., and Thiel, W. R. (2010) Mesoporous organosilicas with acidic frameworks and basic sites in the pores: an approach to cooperative catalytic reactions, *Angew. Chem., Int. Ed.*, **49**, 184–187.
35. Theberge, A. B., Courtois, F., Schaerli, Y., Fischlechner, M., Abell, C., Hollfelder, F., and Huck, W. T. S. (2010) Microdroplets in microfluidics: an evolving platform for discoveries in chemistry and biology, *Angew. Chem., Int. Ed.*, **49**, 5846–5868.
36. Huebner, A., Olguin, L. F., Bratton, D., Whyte, G., Huck, W. T. S., de Mello, A. J., Edel, J. B., Abell, C., and Hollfelder, F. (2008) Development of quantitative cell-based enzyme assays in microdroplets, *Anal. Chem.*, **80**, 3890–3896.
37. Cai, L., Friedman, N., and Xie, X. S. (2006) Stochastic protein expression in individual cells at the single molecule level, *Nature*, **440**, 358–362.
38. Agresti, J. J., Antipov, E., Abate, A. R., Ahn, K., Rowat, A. C., Baret, J. C., Marquez, M., Klibanov, A. M., Griffiths, A. D., and Weitz, D. A. (2010) Ultrahigh-throughput screening in drop-based microfluidics for directed evolution, *Proc. Natl. Acad. Sci. U. S. A.*, **107**, 4004–4009.
39. Poe, S. L., Kobašlija, M., and McQuade, D. T. (2006) Microcapsule enabled multicatalyst system, *J. Am. Chem. Soc.*, **128**, 15586–15587.
40. Poe, S. L., Kobašlija, M., and McQuade, D. T. (2007) Mechanism and application of a microcapsule enabled multicatalyst reaction, *J. Am. Chem. Soc.*, **129**, 9216–9221.
41. Helms, B., Guillaudeu, S. J., Xie, Y., McMurdo, M., Hawker, C. J., and Fréchet, J. M. J. (2005) One-pot reaction cascades using star polymers with core-confined catalysts, *Angew. Chem., Int. Ed.*, **44**, 6384–6387.

42. O'Reilly, R. K., Hawker, C. J., and Wooley, K. L. (2006) Cross-linked block copolymer micelles: functional nanostructures of great potential and versatility, *Chem. Soc. Rev.*, **35**, 1068–1083.

43. Persigehl, P., Jordan, R., and Nuyken, O. (2000) Functionalization of amphiphilic poly(2-oxazoline) block copolymers: a novel class of macroligands for micellar catalysis, *Macromolecules*, **33**, 6977–6981.

44. Kobayashi, S., Wakabayashi, T., Nagayama, S., and Oyamada, H. (1997) Lewis acid catalysis in micellar systems. Sc(OTf)(3)-catalyzed aqueous aldol reactions of silyl enol ethers with aldehydes in the presence of a surfactant, *Tetrahedron Lett.*, **38**, 4559–4562.

45. Chen, H., Li, Y. Z., Chen, J. R., Cheng, P. M., He, Y. E., and Li, H. J. (1999) Micellar effect in high olefin hydroformylation catalyzed by water-soluble rhodium complex, *J. Mol. Catal. A: Chem.*, **149**, 1–6.

46. Lu, A., Cotanda, P., Patterson, J. P., Longbottom, D. A., and O'Reilly, R. K. (2012) Aldol reactions catalyzed by L-proline functionalized polymeric nanoreactors in water, *Chem. Commun.*, **48**, 9699–9701.

47. Minkler, S. R. K., Lipshutz, B. H., and Krause, N. (2011) Gold catalysis in micellar systems, *Angew. Chem., Int. Ed.*, **50**, 7820–7823.

48. Lipshutz, B. H., Ghorai, S., and Aguinaldo, G. T. (2008) Ring-closing metathesis at room temperature within nanometer micelles using water as the only solvent, *Adv. Synth. Catal.*, **350**, 953–956.

49. Lipshutz, B. H., Chung, D. W., and Rich, B. (2008) Sonogashira couplings of aryl bromides: room temperature, water only, no copper, *Org. Lett.*, **10**, 3793–3796.

50. Tong, W. J., Song, X. X., and Gao, C. Y. (2012) Layer-by-layer assembly of microcapsules and their biomedical applications, *Chem. Soc. Rev.*, **41**, 6103–6124.

51. Antipov, A. A., and Sukhorukov, G. B. (2004) Polyelectrolyte multilayer capsules as vehicles with tunable permeability, *Adv. Colloid Interface Sci.*, **111**, 49–61.

52. Möhwald, H. (2000) From Langmuir monolayers to nanocapsules, *Colloid Surf., A*, **171**, 25–31.

53. Peyratout, C. S., and Dähne, L. (2004) Tailor-made polyelectrolyte microcapsules: from multilayers to smart containers, *Angew. Chem., Int. Ed.*, **43**, 3762–3783.

54. Goldberg, M., Langer, R., and Jia, X. Q. (2007) Nanostructured materials for applications in drug delivery and tissue engineering, *J. Biomater. Sci., Polym. Ed.*, **18**, 241–268.

55. Tiourina, O. P., Antipov, A. A., Sukhorukov, G. B., Larionova, N. L., Lvov, Y., and Möhwald, H. (2001) Entrapment of alpha-chymotrypsin into hollow polyelectrolyte microcapsules, *Macromol. Biosci.*, **1**, 209–214.

56. Lvov, Y., Antipov, A. A., Mamedov, A., Möhwald, H., and Sukhorukov, G. B. (2001) Urease encapsulation in nanoorganized microshells, *Nano Lett.*, **1**, 125–128.

57. Fois, A. F., Yap, A., Masters, A. F., and Maschmeyer, T. (2008) Catalyst compartmentalization in dynamic kinetic resolutions, *Top. Catal.*, **48**, 153–157.

58. Qi, W., Yan, X. H., Juan, L., Cui, Y., Yang, Y., and Li, J. B. (2009) Glucose-sensitive microcapsules from glutaraldehyde cross-linked hemoglobin and glucose oxidase, *Biomacromolecules,* **10**, 1212–1216.

59. Qi, W., Yan, X. H., Fei, J. B., Wang, A. H., Cui, Y., and Li, J. B. (2009) Triggered release of insulin from glucose-sensitive enzyme multilayer shells, *Biomaterials,* **30**, 2799–2806.

60. Mak, W. C., Cheung, K. Y., and Trau, D. (2008) Diffusion controlled and temperature stable microcapsule reaction compartments for high-throughput microcapsule-PCR, *Adv. Funct. Mater.,* **18**, 2930–2937.

61. Price, A. D., Zelikin, A. N., Wark, K. L., and Caruso, F. (2010) A biomolecular "ship-in-a-bottle": continuous RNA synthesis within hollow polymer hydrogel assemblies, *Adv. Mater.*, **22**, 720–723.

62. Price, A. D., Zelikin, A. N., Wang, Y. J., and Caruso, F. (2009) Triggered enzymatic degradation of DNA within selectively permeable polymer capsule microreactors, *Angew. Chem., Int. Ed.*, **48**, 329–332.

63. Duan, L., He, Q., Wang, K. W., Yan, X. H., Cui, Y., Möhwald, H., and Li, J. B. (2007) Adenosine triphosphate biosynthesis catalyzed by F0F1 ATP synthase assembled in polymer microcapsules, *Angew. Chem., Int. Ed.,* **46**, 6996–7000.

64. He, Q., Duan, L., Qi, W., Wang, K. W., Cui, Y., Yan, X. H., and Li, J. B. (2008) Microcapsules containing a biomolecular motor for ATP biosynthesis, *Adv. Mater.,* **20**, 2933–2937.

65. Kreft, O., Prevot, M., Möhwald, H., and Sukhorukov, G. B. (2007) Shell-in-shell microcapsules: a novel tool for integrated, spatially confined enzymatic reactions, *Angew. Chem., Int. Ed.*, **46**, 5605–5608.

66. Bäumler, H., and Georgieva, R. (2010) Coupled enzyme reactions in multicompartment microparticles, *Biomacromolecules,* **11**, 1480–1487.

67. Lee, J., Park, J. C., and Song, H. (2008) A nanoreactor framework of a Au@SiO2 yolk/shell structure for catalytic reduction of p-nitrophenol, *Adv. Mater.*, **20**, 1523–1528.

68. Yang, Y., Liu, X., Li, X., Zhao, J., Bai, S., Liu, J., and Yang, Q. (2012) A yolk–shell nanoreactor with a basic core and an acidic shell for cascade reactions, *Angew. Chem., Int. Ed.*, **51**, 9164–9168.

69. Luisi, P., Ferri, F., and Stano, P. (2006) Approaches to semi-synthetic minimal cells: a review, *Naturwissenschaften*, **93**, 1–13.

70. Walde, P., and Ichikawa, S. (2001) Enzymes inside lipid vesicles: preparation, reactivity and applications, *Biomol. Eng.*, **18**, 143–177.

71. Fricker, G., Kromp, T., Wendel, A., Blume, A., Zirkel, J., Rebmann, H., Setzer, C., Quinkert, R. O., Martin, F., and Müller-Goymann, C. (2010) Phospholipids and lipid-based formulations in oral drug delivery, *Pharm. Res.*, **27**, 1469–1486.

72. Steinberg-Yfrach, G., Rigaud, J. L., Durantini, E. N., Moore, A. L., Gust, and D., Moore, T. A. (1998) Light-driven production of ATP catalysed by F0F1-ATP synthase in an artificial photosynthetic membrane, *Nature*, **392**, 479–482.

73. Stano, P., Carrara, P., Kuruma, Y., de Souza, T. P., and Luisi, P. L. (2011) Compartmentalized reactions as a case of soft-matter biotechnology: synthesis of proteins and nucleic acids inside lipid vesicles. J. *Mater. Chem.*, **21**, 18887–18902.

74. Ishikawa, K., Sato, K., Shima, Y., Urabe, I., and Yomo, T. (2004) Expression of a cascading genetic network within liposomes, *FEBS Lett.*, **576**, 387–390.

75. Noireaux, V., and Libchaber, A. (2004) A vesicle bioreactor as a step toward an artificial cell assembly, *Proc. Natl. Acad. Sci. U. S. A.*, **101**, 17669–17674.

76. Wong, B., Boyer, C., Steinbeck, C., Peters, D., Schmidt, J., van Zanten, R., Chmelka, B., and Zasadzinski, J. A. (2011) Design and in situ characterization of lipid containers with enhanced drug retention, *Adv. Mater.*, **23**, 2320–2325.

77. Städler, B., Chandrawati, R., Price, A. D., Chong, S., Breheney, K., Postma, A., Connal, L. A., Zelikin, A. N., and Caruso, F. (2009) A microreactor with thousands of subcompartments: enzyme-loaded liposomes within polymer capsules, *Angew. Chem., Int. Ed.*, **48**, 4359–4362.

78. Discher, B. M., Won, Y. Y., Ege, D. S., Lee, J. C. M., Bates, F. S., Discher, D. E., and Hammer, D. A. (1999) Polymersomes: tough vesicles made from diblock copolymers, *Science*, **284**, 1143–1146.

79. Discher, D. E., and Eisenberg, A. (2002) Polymer vesicles, *Science*, **297**, 967–973.

80. Le Meins, J. , Sandre, O., and Lecommandoux, S. (2011) Recent trends in the tuning of polymersomes' membrane properties, *Eur. Phys. J. E.*, **34**, 1–17.

81. Brinkhuis, R. P., Rutjes, F. P. J. T., and van Hest, J. C. M. (2011) Polymeric vesicles in biomedical applications, *Polym. Chem.*, **2**, 1449–1462.

82. Discher, D. E., and Ahmed, F. (2006) Polymersomes, *Annu. Rev. Biomed. Eng.*, **8**, 323–341.

83. Chen, Q., Schönherr, H., and Vancso, G. J. (2009) Block-copolymer vesicles as nanoreactors for enzymatic reactions, *Small*, **5**, 1436–1445.

84. Chen, Q., Rausch, K. G., Schönherr, H., and Vancso, G. J. (2010) Alpha-chymotrypsin-catalyzed reaction confined in block-copolymer vesicles, *ChemPhysChem*, **11**, 3534–3540.

85. Axthelm, F., Casse, O., Koppenol, W. H., Nauser, T., Meier, W., and Palivan, C. G. (2008) Antioxidant nanoreactor based on superoxide dismutase encapsulated in superoxide-permeable vesicles, *J. Phys. Chem. B*, **112**, 8211–8217.

86. Sauer, M., Haefele, T., Graff, A., Nardin, C., and Meier, W. (2001) Ion-carrier controlled precipitation of calcium phosphate in giant ABA triblock copolymer vesicles, *Chem. Commun.*, 2452–2453.

87. Nardin, C., Thoeni, S., Widmer, J., Winterhalter, and M., Meier, W. (2000) Nanoreactors based on (polymerized) ABA-triblock copolymer vesicles, *Chem. Commun.*, 1433–1434.

88. Nardin, C., Widmer, J., Winterhalter, M., and Meier, W. (2001) Amphiphilic block copolymer nanocontainers as bioreactors, *Eur. Phys. J. E: Soft Matter Bio. Phys.*, **4**, 403–410.

89. Ranquin, A., Versées, W., Meier, W., Steyaert, J., and Van Gelder, P. (2005) Therapeutic nanoreactors: combining chemistry and biology in a novel triblock copolymer drug delivery system, *Nano Lett.*, **5**, 2220–2224.

90. Nallani, M., Benito, S., Onaca, O., Graff, A., Lindemann, M., Winterhalter, M., Meier, W., and Schwaneberg, U. (2006) A nanocompartment system (Synthosome) designed for biotechnological applications, *J. Biotechnol.*, **123**, 50–59.

91. Kim, K. T., Cornelissen, J. J. L. M., Nolte, R. J. M., and van Hest, J. C. M. (2009) A polymersome nanoreactor with controllable permeability induced by stimuli-responsive block copolymers, *Adv. Mater.*, **21**, 2787–2791.

92. Gaitzsch, J., Appelhans, D., Wang, L. G., Battaglia, and G., Voit, B. (2012) Synthetic Bio-nanoreactor: mechanical and chemical control

of polymersome membrane permeability, *Angew. Chem., Int. Ed.*, **51**, 4448–4451.

93. Wang, L. G., Chierico, L., Little, D., Patikarnmonthon, N., Yang, Z., Azzouz, M., Madsen, J., Armes, S. P., and Battaglia, G. (2012) Encapsulation of biomacromolecules within polymersomes by electroporation, *Angew. Chem., Int. Ed.*, **51**, 11122–11125.

94. Ha, W., Meng, X. W., Li, Q., Fan, M. M., Peng, S. L., Ding, L. S., Tian, X., Zhang, S., and Li, B. J. (2010) Self-assembly hollow nanosphere for enzyme encapsulation, *Soft Matter*, **6**, 1405–1408.

95. Vriezema, D. M., Hoogboom, J., Velonia, K., Takazawa, K., Christianen, P. C. M., Maan, J. C., Rowan, A. E., and Nolte, R. J. M. (2003) Vesicles and polymerized vesicles from thiophene-containing rod–coil block copolymers, *Angew. Chem., Int. Ed.*, **42**, 772–776.

96. Nallani, M., de Hoog, H. P. M., Cornelissen, J. J. L. M., Palmans, A. R. A., van Hest, J. C. M., and Nolte, R. J. M. (2007) Polymersome nanoreactors for enzymatic ring-opening polymerization, *Biomacromolecules*, **8**, 3723–3728.

97. Kuiper, S. M., Nallani, M., Vriezema, D. M., Cornelissen, J. J. L. M., van Hest, J. C. M., Nolte, R. J. M., and Rowan, A. E. (2008) Enzymes containing porous polymersomes as nano reaction vessels for cascade reactions, *Org. Biomol. Chem.*, **6**, 4315–4318.

98. Vriezema, D. M., Garcia, P. M. L., Oltra, N. S., Hatzakis, N. S., Kuiper, S. M., Nolte, R. J. M., Rowan, A. E., and van Hest, J. C. M. (2007) Positional assembly of enzymes in polymersome nanoreactors for cascade reactions, *Angew. Chem., Int. Ed.*, **46**, 7378–7382.

99. van Dongen, S. F. M., Nallani, M., Cornelissen, J. J. L. M., Nolte, R. J. M., and van Hest, J. C. M. (2009) A three-enzyme cascade reaction through positional assembly of enzymes in a polymersome nanoreactor, *Chem. Eur. J.*, **15**, 1107–1114.

100. Meeuwissen, S. A., Rioz-Martínez, A., de Gonzalo, G., Fraaije, M. W., Gotor, V., and van Hest, J. C. M. (2011) Cofactor regeneration in polymersome nanoreactors: enzymatically catalysed Baeyer-Villiger reactions, *J. Mater. Chem.*, **21**, 18923–18926.

101. Choi, H. J., and Montemagno, C. D. (2005) Artificial organelle: ATP synthesis from cellular mimetic polymersomes, *Nano Lett.*, **5**, 2538–2542.

102. Marguet, M., Bonduelle, C., and Lecommandoux, S. (2013) Multicompartmentalized polymeric systems: towards biomimetic cellular structure and function, *Chem. Soc. Rev.*, **42**, 512–529.

103. Fu, Z. K., Ochsner, M. A., de Hoog, H. P. M., Tomczak, N., and Nallani, M. (2011) Multicompartmentalized polymersomes for selective encapsulation of biomacromolecules, *Chem. Commun.*, **47**, 2862–2864.

104. Marguet, M., Edembe, L., and Lecommandoux, S. (2012) Polymersomes in polymersomes: multiple loading and permeability control, *Angew. Chem., Int. Ed.*, **51**, 1173–1176.

105. Shum, H. C., and Weitz, D. A. (2011) Multicompartment polymersome gel for encapsulation, *Soft Matter*, **7**, 8762–8765.

106. Shum, H. C., Zhao, Y. J., Kim, S. H., and Weitz, D. A. (2011) Multicompartment polymersomes from double emulsions, *Angew. Chem., Int. Ed.*, **50**, 1648–1651.

107. Kim, S., Shum, H. C., Kim, J. W., Cho, J., and Weitz, D. A. (2011) Multiple polymersomes for programmed release of multiple components, *J. Am. Chem. Soc.*, **133**, 15165–15171.

108. Thompson, K. L., Chambon, P., Verber, R., and Armes, S. P. (2012) Can polymersomes form colloidosomes? *J. Am. Chem. Soc.*, **134**, 12450–12453.

109. Wang, Z. P., van Oers, M. C. M., Rutjes, F. P. J. T., and van Hest, J. C. M. (2012) Polymersome colloidosomes for enzyme catalysis in a biphasic system, *Angew. Chem., Int. Ed.*, **51**, 10746–10750.

110. Cardinale, D., Carette, N., and Michon, T. (2012) Virus scaffolds as enzyme nano-carriers, *Trends Biotechnol.*, **30**, 369–376.

111. Comellas-Aragonès, M., Engelkamp, H., Claessen, V. I., Sommerdijk, N. A. J. M., Rowan, A. E., Christianen, P. C. M., Maan, J. C., Verduin, B. J. M., Cornelissen, J. J. L. M., and Nolte, R. J. M. (2007) A virus-based single-enzyme nanoreactor, *Nat. Nanotechnol.*, **2**, 635–639.

112. Bode, S. A., Minten, I. J., Nolte, R. J. M., and Cornelissen, J. J. L. M. (2011) Reactions inside nanoscale protein cages, *Nanoscale*, **3**, 2376–2389.

113. Minten, I. J., Claessen, V. I., Blank, K., Rowan, A. E., Nolte, R. J. M., and Cornelissen, J. J. L. M. (2011) Catalytic capsids: the art of confinement, *Chem. Sci.*, **2**, 358–362.

114. Inoue, T., Kawano, M. A., Takahashi, R. U., Tsukarnoto, H., Enornoto, T., Imai, T., Kataoka, K., and Handa, H. (2008) Engineering of SV40-based nano-capsules for delivery of heterologous proteins as fusions with the minor capsid proteins VP2/3, *J. Biotechnol.*, **134**, 181–192.

115. Douglas, T., and Young, M. (1998) Host-guest encapsulation of materials by assembled virus protein cages, *Nature*, **393**, 152–155.

Chapter 5

Nanocarriers for Therapeutic Enzymes

Melissa D. Howard, Elizabeth D. Hood, and Vladimir R. Muzykantov
Department of Pharmacology and Center for Targeted Therapeutics and Translational Nanomedicine, Perelman School of Medicine, University of Pennsylvania, 3400 Civic Center Boulevard, Philadelphia, PA 19104, USA
muzykant@mail.med.upenn.edu

5.1 Introduction

Enzymes have tremendous potential to serve as highly efficacious drugs with limited off-target effects in a plethora of disease conditions, including ischemia, bleeding, thrombosis, inflammatory disorders, cancer, and lysosomal storage disorders (LSDs), among others. Compared to traditional small-molecule drugs, these biotherapeutics influence their molecular targets with exceptional specificity and with high potency and efficacy due to their catalytic nature. Enzymes, however, are costly, labile, often have unfavorable pharmacokinetics (PK) and usually require delivery to specific compartments in the body, sometimes on a subcellular level, to be efficacious. For these reasons, nanocarriers (NCs) are being explored as enzyme drug delivery vehicles.

Enzyme Nanocarriers
Edited by Daniela Cardinale and Thierry Michon
Copyright © 2016 Pan Stanford Publishing Pte. Ltd.
ISBN 978-981-4613-42-2 (Hardcover), 978-981-4613-43-9 (eBook)
www.panstanford.com

Although untargeted NCs acting via passive uptake have been studied extensively, these systems are subject to limitations that may make them unsuitable for directing therapeutic enzymes to their sites of action. Untargeted NCs may enter cells passively depending on their physical characteristics, size, surface charge, or surface composition [1]. For example, cells tend to internalize small (<100 nm diameter) positively charged particles [2, 3]. Additionally, particles may accumulate in tissues (e.g., rapidly forming tumors) characterized by a compromised endothelial barrier structure [4, 5]. The resulting "leaky" vasculature enables the NCs to accumulate to a greater degree than in normal tissues. However, this "enhanced permeability and retention" (EPR) effect does not have the capacity to direct carriers to specific cellular or subcellular sites of action. Active delivery of therapeutic NCs through the use of targeting ligands (monoclonal antibodies, their affinity by-product single chains or fusion proteins, peptides) enables tissue and cell-specific localization as well as control of uptake pathways and subsequently may direct the subcellular fate of the therapeutic cargos. We will address the use of targeting to cellular adhesion molecules (CAMs) expressed on the surface of endothelial cells.

In this chapter, we review this research area in the context of vascular pathology and drug delivery to distinct therapeutic targets. They include endothelial cells lining the vessel lumen and exerting important systemic regulatory functions, components of blood clots, immune cells, and intracellular compartments of metabolically challenged cells. Specifically, we analyze the state of the art in NCs developed for diverse classes of therapeutic enzymes (fibrinolytics, antioxidant enzymes [AOEs], enzymes for antitumor therapy, detoxifying enzymes, and enzymes for enzyme replacement therapy [ERT]).

5.2 Nanocarriers

5.2.1 Rationale for the Use of Nanocarriers

Enzymes injected in their native state may be rapidly cleared from circulation due to proteolytic degradation, renal filtration, and/or clearance by the reticuloendothelial system (RES). They also usually fail to reach the compartments necessary for interaction with their therapeutic targets.

NCs have the potential to overcome these challenges of enzyme therapy. Conventionally, NCs consist of a vesicle or solid core delivery unit that carries the loaded drug cargo, thereby improving its PK and limiting its interactions with the body en route. Newer generations of NCs have been designed with multifunctional elements to provide targeted delivery, bioresponsive drug release, and enhanced membrane permeation to improve drug localization and timing (Fig. 5.1) [6].

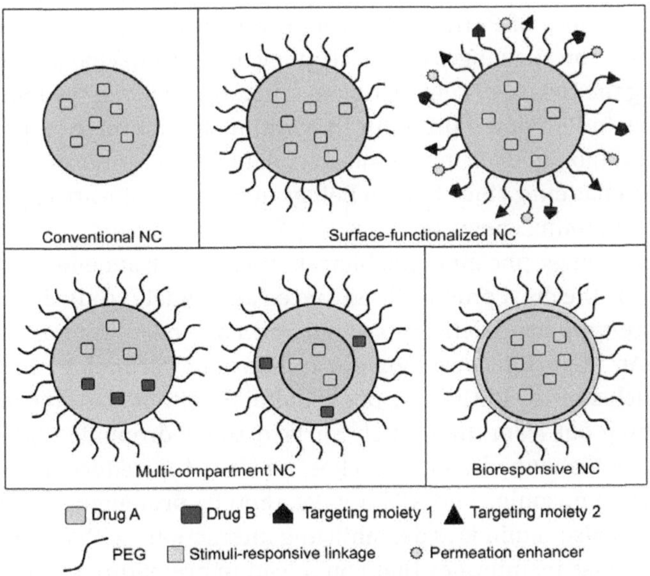

Figure 5.1 Multifunctional nanocarrier (NC) design. Enzymes can be loaded into conventional NCs (top left), NCs surface-functionalized with PEG and/or various targeting agents and permeation enhancers (top right), NCs with multiple compartments for loading of drugs with different properties (bottom left), or NCs designed to release drug in a bioresponsive manner (bottom right).

5.2.2 Types of Nanocarriers

A wide variety of NCs have been devised for enzyme delivery. Broadly, this list consists of polymer–enzyme conjugates, conventional and ultrasound-sensitive (echogenic) liposomes, dendrimers, and polymeric, perfluorocarbon (PFC), and magnetic nanoparticles (MNPs). Within each subset, the use of different materials, linkers, and surface coatings allows for further optimization of the NC size, geometry, stability, drug loading, and release kinetics.

One of the first modifications of therapeutic proteins developed and translated into clinical use was covalent conjugation with hydrophilic polymers, specifically poly(ethylene glycol) (PEG). PEGylation of enzymes restricts their renal filtration, cellular uptake, proteolytic degradation, and immunogenicity ("stealth" technology) [7]. A number of PEGylated enzymes are currently in clinical trials and use. The challenges of this strategy include enzyme inactivation during the course of conjugation and heterogeneity of the molecular species in PEGylated proteins. In addition, the immune response to the PEG moiety itself, negligible in rodents, may represent a serious issue in patients, since human beings in everyday life are exposed to PEG-containing chemicals.

Liposomes, phospholipid bilayer vesicles with aqueous cores, are arguably the most practically advanced and widely studied nanoparticulate-type drug delivery system and have been used as enzyme NCs. An important development in liposome technology has been the inclusion of PEGylated phospholipids within the lipid bilayer. Reaching out from the particle surface, the hydrophilic PEG chains produce highly hydrated branches, which act to reduce adsorption of plasma opsonins and RES uptake, thereby prolonging circulation [8]. This also minimizes destabilizing interactions with lipoproteins and cellular membranes that could lead to premature drug release. Furthermore, the terminal ends of PEG moieties represent advantageous sites for conjugation of affinity moieties (e.g., monoclonal antibodies or peptides). Several liposomal systems are used clinically to reduce the toxicity of chemotherapeutic agents [6] (e.g., Doxil is a liposomal form of doxorubicin [9, 10]).

Echogenic liposomes (ELIPs) are a variation of conventional liposomes. These particles contain both gas and fluid in the core, allowing them to be used for both drug delivery and ultrasound imaging. Similarly, PFC nanoparticles, which are characterized by a liquid PFC core encapsulated by a phospholipid monolayer, can be used for noninvasive imaging using both ultrasound and magnetic resonance modalities [11–13].

MNPs are solid particles or polymorphous complexes formed using diverse materials (polymers, proteins, PEG-containing stabilizers, and stealth moieties) and methods (e.g., controlled precipitation of drugs in multiphase suspensions). Inclusion of magnetically responsive metal nanoparticles (generally, 5–10 nm diameter iron grain) endows formed NCs with the ability to accumulate in certain areas on the basis of the presence of an external magnetic field or magnetized stent [14, 15].

Polymer-based NCs, such as dendrimers and polymeric micelles, have attracted increasing attention due to their capacity for customization. These NCs may be prepared from either synthetic, for example, poly(lactic-co-glycolic acid) (PLGA) or biological polymers, for example, chitosan and poly(nucleic acids), as well as blends of different polymers conjugated into diblocks, triblocks, and higher-order copolymers. For example, dendrimers are repetitively and structurally regularly branched spherical multimolecular complexes, with diameters usually in the range of 10–100 nm. Drugs can be coupled to the active end groups or encapsulated into the meshwork, although the loading capacity is generally lower than for other NCs [16–18].

Polymeric micelles or nanoparticles are typically based on amphiphilic block copolymers, which form a distinct core–corona structure in solution [19]. The core comprises the hydrophobic portion of the polymer, while the corona consists of the hydrophilic portion (e.g., PEG). Due to enhanced stability and stealth features, these carriers may exhibit longer PK and residence time in tissues versus lipid-based carriers. Yet, these particles are challenging for enzyme loading due to the organic solvents and/or sheer involved in their preparation, resulting in destruction of protein mass and activity. Other difficulties include achieving a desirable drug release profile and minimizing toxicity.

The majority of current NCs are spherical. However, recently, there has been increased attention on shape as a design parameter to enhance cellular delivery and prolong circulation time, among other facets. For example, filomicelles represent a variation on classical spherical polymeric micelles. Similarly prepared from amphiphilic block copolymers, these particles assemble in dynamic, flow-responsive filamentous morphologies [20, 21]. By aligning with flow and minimizing interaction with cells, these particles may remain in circulation for up to several days, or 10 times longer than their spherical counterparts [22]. The kinetics of cellular uptake of nonspherical NCs may differ from those of their spherical counterparts, depending on the angle of anchoring on the plasmalemma [23].

5.3 Cargos

5.3.1 Fibrinolytics

Plasminogen activators, known as fibrinolytics, are a class of proteolytic enzymes used to dissolve blood clots. Mechanistically, they work through catalyzing the conversion of plasminogen to plasmin, the enzyme responsible for breakdown of the fibrin in blood clots. Administration during or shortly after a heart attack or stroke may prevent clots from blocking the flow of blood to the heart or brain. Unfortunately, fibrinolytics have unfavorable PK, that is, fast blood clearance and/or inactivation. Further, indiscriminate activation of plasmin can lead to an increased risk of bleeding. NCs may be used to extend the circulation lifetime of the enzymes and potentially achieve a level of targeting specificity.

PEGylated plasminogen activators are the simplest form of fibrinolytic NCs. Since the early 1980s, numerous in vitro and in vivo studies have tested the effects of PEG molecular weight, conjugation chemistry, and extent of coverage on the functional activity and circulation lifetime of tissue plasminogen activator (tPA) [24], streptokinase (SK) [25], urokinase [26, 27], and staphylokinase [28]. Generally, PEGylation has been shown to reduce immune reactions and proteolytic degradation of the proteins [25, 29], and PEGylated fibrinolytics show prolonged circulation and enhanced bioavailability [24–27, 29, 30].

PEGylation is, however, a balancing act between achieving extended PK and losing protein activity due to chemical modification (usually via lysine residues, which may lead to conformational changes) or interference of PEG chains in interaction with plasminogen [24, 26]. Optimized PEGylation of plasminogen activators allows for prolonged PK with minimal loss of activity, resulting in enhanced clot lysis in animal studies [27, 30]. In the pilot clinical trial with PEGylated staphylokinase, plasma half-lives ($t_{1/2}$) of 13±1.5 minutes and 120±8.8 minutes were achieved with proteins modified with PEG chains of 5 K and 20 K molecular weight, respectively, as compared to a $t_{1/2}$ of 3 minutes for the non-PEGylated protein. Out of 18 patients, 14 received a 5 mg bolus dose of the 5 K PEGylated derivative, achieving thrombolysis in myocardial infarction (TIMI) grade 3 flow (grade 3 is defined as normal coronary patency) within 60 minutes that persisted to 90 minutes. A phase IIa clinical trial (CAPTORS II) confirmed that PEGylated staphylokinase had efficacy similar to recombinant tPA, but as side effects were not reduced, no further studies were pursued [28].

While PEGylation significantly prolongs the half-life of fibrinolytics, this improvement has not translated into tangible therapeutic benefits. More sophisticated NCs, such as liposomes, have now been evaluated as alternatives. The first attempts at liposomal delivery of fibrinolytics began in the late 1980s. Similarly to PEG, liposomes protect the enzymes from degradation [31], minimize immune reactions [32], and increase plasma half-life [33, 34]. Liposomal tPA showed enhanced potency in a rabbit jugular vein thrombosis model [35], while liposomal SK also significantly reduced reperfusion time as compared to free SK (19.3±4.6 vs. 74.5±16.9 min, respectively) in a rabbit model of carotid artery thrombosis [36].

Polymer-based NCs have also been studied for delivery of fibrinolytics, though they are generally less well developed than liposomes. Examples include chitosan/PLGA nanoparticles for tPA [37], glycol chitosan/PEG nanogels for urokinase [38], polyglycerol [39], and poly(amido amine) (PAMAM) [40] dendrimers for SK. In the canine model of coronary artery thrombosis, SK encapsulated into poly(vinyl alcohol) (PVA)/PEG particles (~800 nm) achieved reperfusion significantly faster than free SK or SK with control particles (29±4 min vs. 68±7 min and 83±18 min, respectively). Residual clot mass, infarct mass, reocclusion episodes, fibrinogen depletion, and blood loss were also reduced with encapsulated SK as compared to free SK. Despite these relative improvements, clinical translation of these approaches continues to be limited by a number of considerations. For example, permeation into clots is more challenging for NCs than much smaller proteins. Control of timely drug release in the clots is yet another unresolved challenge. In general, stealth strategies, including PEGylated liposomes, prolong circulation of fibrinolytics at par with genetic modifications of the enzymes. However, delivery systems that prolong drug circulation may also cause more profound adverse side effects due to systemic off-target activity.

To localize the effect of fibrinolytics, several methods have been explored, including use of targeting ligands, external triggers, and hemodynamic factors. A number of clot components have been considered as possible targets. Targeting the integrin GPIIb-IIIa on activated platelets can be achieved using sRGD peptide, a fibrinogen mimic [41–43]. In the rat model containing a human blood clot inoculated in the carotid artery, RGD-conjugated liposomes loaded with SK achieved 34% clot lysis 30 minutes following administration as compared to 22% with a free fibrinolytic [42]. Nontargeted liposomes produced intermediate results. P-selectin is also upregulated on activated platelets, making it a second additional target. Liposomes prepared with a glycolipid of 3′-sulfo-Lewis a ($SuLe^a$)-PEG-DSPE have been designed to mimic P-selectin's natural ligand, P-selectin glycoprotein ligand 1 [44]. Targeting particles was confirmed in vitro, but no therapeutic efficacy data was reported.

PFC nanoparticles conjugated with monoclonal antifibrin were also designed to target urokinase to clots; thrombolytic efficacy was dependent on the number of enzymes bound per particle [45]. Fluorescent microscopy showed that rhodamine-labeled urokinase nanoparticles surrounded the intraluminal thrombus in canine clots in vivo analogous to the fibrin pattern, while an irrelevant targeted agent showed negligible binding. Crosslinked dextran-coated iron oxide nanoparticles loaded with tPA have been targeted to activated factor XIIIa, the enzyme responsible for crosslinking fibrin [46]. Microscopy experiments confirmed binding of the targeted particles to the periphery of clots, and thrombolytic efficacy was found to be comparable to free tPA in mice injected with preformed human clots.

External triggers may also be used to improve the targeting capacity of NCs. For example, in the field of liposome research, there has been considerable interest in preparing ultrasound-sensitive ELIPs. Importantly, they may provide a means of targeting drug to the clot, depending on the ability of the ultrasound to (1) accelerate drug release from the ELIPs and (2) enhance drug transport into the thrombi. To date, both in vitro and in vivo results indicate that tPA-loaded ELIPs have lytic efficacy at least comparable to free tPA; ultrasound exposure increases the lysis achieved with both free tPA and tPA-loaded ELIPs [47–49]. Encouraging results were also obtained with an ultrasound-responsive PEG-gelatin NC loaded with tPA [50]. In a rabbit model, tPA circulation was prolonged three times as compared to free tPA, which improved recanalization of occlusive clots. However, efficacy of delivery and ultrasound-mediated inactivation of the loaded enzymes represent difficult to resolve challenges [51, 52].

MNPs may potentially be targeted to the site of a clot through the use of external magnets. Urokinase-conjugated dextran-conjugated MNPs combined with targeting magnets were 5-fold more efficient at achieving thrombolysis than urokinase and 2.6-fold more efficient than urokinase-conjugated MNPs without targeting magnets in a rat arteriovenous shunt thrombosis model [53]. Also, tail-bleeding times were insignificantly increased with targeted urokinase-conjugated MNPs treatment, whereas administration of free urokinase or nontargeted urokinase-conjugated particles resulted in at least two-fold extended bleeding times. In a rat embolic model, poly(acrylic acid)-coated MNPs loaded with tPA-restored iliac blood flow within 75 minutes to 82% of that before clot lodging when combined with targeting magnets [15]. An equivalent amount of free tPA resulted in no improvement in flow. At the end of a two-hour period, tPA-loaded MNPs did not alter the levels of hemoglobin, hematocrit, or blood cell count, demonstrating their localized activity.

A unique approach proposed by Absar et al. combines both clot and external targeting [54]. Nanoparticles are formed by complexing a tPA–low-molecular-weight heparin conjugate with an albumin–protamine conjugate. In this form, tPA remains inactive, until an excess of free heparin releases it from the complex. Coating the particles with RGD peptide enables their binding to activated platelets. In theory, heparin administration can cause release of tPA, leading to degradation of the clots. Studies in a rat jugular vein thrombosis model showed that the clot lysis of the heparin-triggered camouflaged tPA group was equivalent to the tPA + heparin group without prolongation of activated partial thromboplastin time. However, the complexity of this double conjugate, uncertainty for timing of heparin injection in real clinical settings, and side effects of heparin complicate the development and utility of this delivery system.

Most recently, studies have focused on achieving targeting through taking advantage of the elevated shear stress present around blood clots. The Ingber group has reported on the development of shear-activated nanotherapeutics (SA-NTs), micron-sized aggregates (3.8±1.6 μm) composed of multiple smaller tPA-coated nanoparticles (180±70 nm) [55]. The aggregates remain intact under normal physiological flow conditions (10–30 dyne/cm^2) but are triggered to break up into their individual components upon exposure to the elevated shear stress encountered in the vicinity of blood clots (>100 dyne/cm^2, comparable to a 60% lumen obstruction). Because the nanoparticles are smaller and experience lower drag forces than the microparticles, they can adhere more efficiently to the damaged blood vessel, generating a region of high drug concentration (Fig. 5.2). In a mouse arterial thrombosis model, these particles extended the time to vessel occlusion from 12±3 minutes with PBS to 29±7 minutes. SA-NTs were also shown to clear pulmonary emboli in an ex vivo model at a 100-fold lower dose than free tPA. Lenticular lipid-based vesicles that show shear-stress induced drug release have also been reported but have not yet been evaluated for fibrinolytic delivery [56].

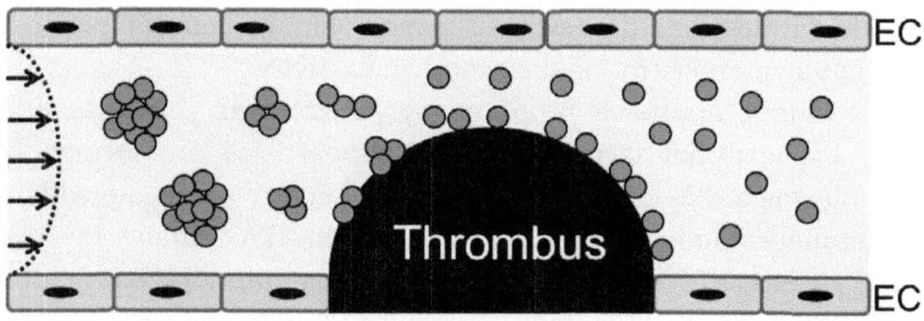

Figure 5.2 Shear-activated nanotherapeutics (SA-NTs) are triggered to break up into individual nanoparticles upon exposure to the elevated shear stress encountered in the vicinity of blood clots. Because the nanoparticles are smaller and experience lower drag forces than the microparticles, they can adhere more efficiently to the damaged blood vessel, generating a region of high drug concentration. EC, endothelial cells.

NCs designed to prolong the circulation and improve delivery of fibrinolytics are not yet translated into clinical practice. Although each carrier has a distinct set of advantages, none have prolonged the circulation of fibrinolytics to the extent necessary to capitalize on the dynamic nature of developing clots. The impermeability of maturing clots further limits their therapeutic utility. Use of external targeting methods may find applicability in situations where the clot location is known and accessible, though these situations may be limited. Targeted and flow-sensitive NCs also may be useful in settings characterized by partial thrombotic occlusion and pathologically high shear stress, although their utility in low shear conditions may be more limited.

5.3.2 Antioxidant Enzymes

AOEs, such as superoxide dismutase (SOD) and catalase that decompose superoxide and H_2O_2, respectively, are more potent than nonenzymatic antioxidant molecules that are consumed in reactions with reactive oxygen species (ROS) [57]. AOE delivery to sites of ROS production, however, is problematic and may benefit from using NCs. Early in vitro studies on liposomal delivery of SOD showed a 6–12-fold increase in activity and a decreased sensitivity to hyperoxia compared to cells treated with free SOD [58]. Hypertension mediated through angiotensin II activation of NADPH oxidase was also reduced in rats treated with liposomal SOD [59].

Early success with untargeted antioxidant liposomes provided the motivation for subsequent improvements including PEG-stealth liposomes, vascular immunotargeting, and advanced NC design. A study examining the effect of synthesis methods on loading efficiency and efficacy showed that acylated superoxide dismutase (Ac-SOD) liposomes formed by the dehydration–rehydration method without extrusion yielded the greatest loading efficiency of active SOD and inhibited oxidative edema in animal models more effectively than naked SOD [60]. Since the Ac-SOD localized to the bilayer and 50% of the enzyme was exposed to the exterior, the site of activity was focused at the surface of the liposome or "enzymosome," instead of within the aqueous interior. The change in conformation reduced the effect of release rate on the activity of the liposome and increased bioavailability of the enzyme [61]. Correspondingly, the SOD location in the liposome structure was shown to increase the anti-inflammatory effect of PEGylated Ac-SOD liposomes in a rat adjuvant rat arthritis model compared to plain SOD PEG liposomes [62]. The Burgess group investigated the incorporation of SOD into liposomes during the latter freeze–thaw stage of formulation, reporting a 50% entrapment efficiency and high retention of activity compared to addition at the thin-film hydration step [63].

Catalase electrostatically complexed with the cationic block copolymer, polyethyleneimine-poly(ethylene glycol) (PEI-PEG), termed a "nanozyme," was delivered to inflamed mouse brains in a model of Parkinson's disease, following macrophage uptake [64]. By both morphological and biochemical measures of brain injury, the delivery system showed promising therapeutic efficacy [65]. Similarly formed SOD nanozymes alleviated neuronal oxidative stress after local administration in the central nervous system (CNS) in rodents [66]. There are reports that conjugation with PEG-containing pluronics increases drug permeation through biological barriers in cell culture [67] and in animals [67, 68]. Although mechanisms of intracellular transport remain undefined, there have been no deleterious effects reported [69]. SOD–pluronic conjugates were reported to deliver enzymatically active SOD to neuronal cells more effectively than naked SOD or PEG-SOD without neuronal toxicity [70]. The mechanism and efficacy of delivery and effect of this antioxidant NC formulation, apparently dependent on somewhat enigmatic features of the migrating phagocytes providing a ride for the drug, remain to be better understood.

SOD covalently functionalized with vinyl groups then followed by free-radical polymerization in the presence of other diacrylate monomers, resulted in encapsulation of the protein molecule in a nanometer-thick (~5 nm) polymer shell [71]. The surface properties and degradation rate of the polymer shell were controlled by monomer selection. SOD nanocapsules prevented cell death in an in vitro paraquat-induced oxidative injury model, suggesting that their polymer shell is permeable to the substrate molecules (for example, $O_2^{\cdot -}$) [71].

MNPs have also been used to load either active SOD or catalase through a controlled precipitation method. Complexation of AOE into MNPs provides advantages over other more traditional carriers that involve the use of either strong sheer or organic solvents, both of which are damaging to a large AOE. The formulation loaded active enzyme cargo efficiently and protected it from proteolytic degradation in cultured endothelial cells, while inhibiting release of the enzyme in plasma. Additionally, magnetically delivered, catalase-containing MNPs protected endothelial cells in vitro, with 62±12% cells rescued from H_2O_2-induced cell death versus 10±4% under nonmagnetic conditions [72].

Encapsulation of AOEs in a polymer shell can also provide protection from proteolytic inactivation, thereby extending therapeutic duration. Using PEG as a stabilizer enhances the efficiency of encapsulation of SOD or catalase into polymeric carriers [73]. By introducing a freezing step during the primary emulsification enzyme inactivation was minimized [74, 75]. When using an amphiphilic encapsulating polymer (e.g., PEG-PLGA), polymer NCs produced through this method were 200–300 nm in size and contained 10 times the mass of catalase compared to synthesis without the cryogenic step. These carriers maintained ~25% of the original catalase activity after 24 hours incubation with a proteolytic enzyme, while free enzyme lost >90% activity within the first hour [74].

Targeting of enzyme carriers may further improve therapeutic effects. For example, cell adhesion molecules are expressed on endothelial cells constitutively, and their surface expression is elevated in many pathological conditions. Conjugation with antibodies to intercellular adhesion molecules (ICAMs) and platelet endothelial cell adhesion molecules (PECAMs) enables endothelial delivery of enzymes and NCs (Fig. 5.3). For instance, PECAM antibodies (anti-PECAM) were coupled to PEG-PLGA NCs loaded with catalase, providing delivery of 12% of the injected dose to the pulmonary vasculature after intravenous injection (lung is a preferred vascular target) as compared to 2% for nontargeted control NCs. As a result, anti-PECAM-coated NCs loaded with catalase, but not non-targeted counterparts, inhibited the ROS levels in the lungs [75].

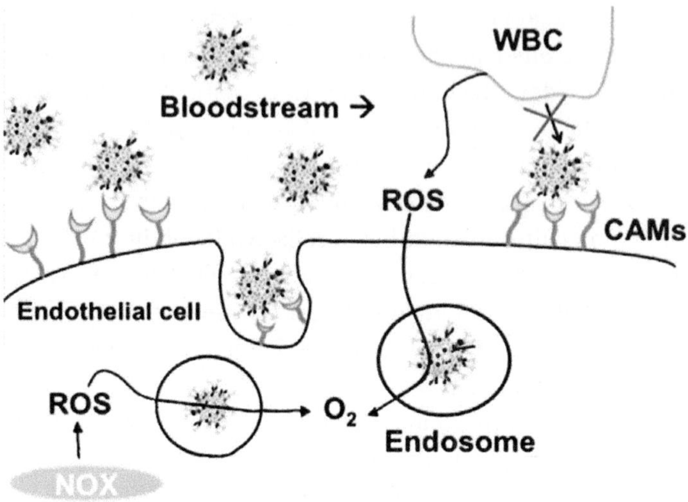

Figure 5.3 Subcellular delivery of AOEs by NCs leads to the reduction of reactive oxygen species. Conjugation of NCs with antibodies to intercellular adhesion molecules (ICAMs) and platelet endothelial cell adhesion molecules (PECAMs) allows for enhanced binding to and internalization by endothelial cells as compared to nontargeted NCs.

No clinically proven AOE carrier has yet entered the medical domain. Nevertheless, the impressive progress achieved in preclinical studies and the notion that AOEs represent relatively benign cargos provide encouragement for the ultimate success of this strategy.

5.3.3 Enzymes for Antitumor Therapy

Some enzymatic reactions in the body destroy molecules needed for active growth of the cells. Depriving tumors of these nutrients or growth-stimulating molecules may help control malignancies. For example, L-asparaginase (ASNase), the enzyme catalyzing the hydrolysis of L-asparagine into L-aspartic acid and ammonia, has found unique use as a tumor inhibitory enzyme in children with acute lymphoblastic leukemia (ALL) [76]. Asparagine is an essential nutrient to leukemic blasts, and they lack the enzyme necessary for its synthesis (asparagine synthetase). Its removal from circulating blood pools can lead to starvation and death of the tumor. Beneficially, normal cells have the ability to synthesize ASN and are thus unaffected. Due to the short circulation lifetime of the native enzyme ($t_{1/2}$ = 1.1 days), a PEGylated version with a $t_{1/2}$ of ~6 days was developed [76]. Today, both are in common clinical use (*E. coli* native ASNase, Merck & Co, Inc.; PEG-ASNase, ONCOSPAR®, Enzon Pharmaceuticals).

Several different types of nanoparticles have also been studied as potential carriers for ASNase. The enzyme was initially loaded in liposomes with good encapsulation efficiency (40% in extruded vesicles <250 nm) and preservation of enzymatic activity (74%–98% of loaded) [77]. Loading in liposomes prolonged the circulation time by a factor of 10 and prevented the induction of antibodies against the enzyme [78]. Further, the survival of tumor-bearing animals was prolonged by a factor of 2 after treatment with encapsulated enzyme as compared to the free enzyme [78]. However, to potentially achieve a more prolonged circulation and drug release, the same group pursued polymer NCs. When formulated in poly(lactide-co-glycolide) (PLG) nanoparticles, encapsulation efficiency and release were shown to be highly dependent on both the PLG molecular weight and the presence of free carboxyl end groups in the PLG chain [78]. Nanoparticles with free carboxyl end groups could provide continuous enzyme delivery for 28 days. Most recently, ASNase has been loaded into dendritic polyglycerol nanogels through an inverse nanoprecipitation process [79]. With this process, encapsulation efficiency was nearly 100% and enzymatic activity was fully preserved. Further studies are needed to evaluate the therapeutic efficacy of these newly developed particles.

Arginine deiminase (ADI) has found similar use in treating cancer, depending on its degradation of arginine, an amino acid necessary for cell growth and development. Because many cancer cells lack argininosuccinatesynthetase (ASS), the enzyme responsible for the synthesis of arginine, depletion of external arginine leads to reduced tumor growth. ADI conjugated to a 20 kD PEG chain (ADI-PEG) has entered phase I/II clinical trials in hepatocellular carcinoma [80, 81] and melanoma [82, 83]. In patients with metastatic melanoma, it was confirmed that a dose of 160 U/m^2 lowered plasma arginine from a resting level of approximately 130 μmol/L to less than 2 μmol/L for at least seven days [82]. Out of 24 phase I to II patients, 6 responded to the treatment (five partial responses and one complete response;

25% response rate) and had prolonged survival. Although no complete responses were observed in Asian advanced hepatocellular carcinoma patients treated with a dose of 160 U/m^2, 31% exhibited stable disease [81]. Recent studies have begun to consider ADI-PEG as a potential therapy in other diseases, including ulcerative colitis [84] and hepatitis C [85].

On the basis of the clinical results of these PEGylated enzymes, prolonging the circulation of enzymes for antitumor therapy is a promising strategy. It remains to be seen whether encapsulation in NCs results in sufficiently improved efficacy to warrant increased complexity.

5.3.4 Detoxifying Enzymes

Detoxifying enzymes also may benefit from delivery via NCs. These enzymes are responsible for the hydrolysis of a wide variety of toxic compounds within the body. Correspondingly, they have been found to be beneficial in a number of disease states.

For example, uricase is an enzyme involved in the purine degradation pathway and is responsible for catalyzing the hydrolysis of uric acid to allantoin. In gout, elevation of uric acid level in blood eventually leads to deposition of uric acid crystals in the joints, causing a painful form of inflammatory arthritis. Studies with PEGylated uricase began in the late 1970s [86, 87]. Finally, in 2010, a 10 kDa mPEG-modified porcine-like recombinant uricase (Pegloticase, Krystexxa®, Savient Pharmaceuticals) was approved by the US Food and Drug Administration for the treatment of gout in adult patients refractory to conventional therapy [88]. The $t_{1/2}$ for the PEGylated enzyme was shown to be in the range of 6.4–13.8 days, and with doses of 4–12 mg, the plasma urate concentration fell from 11.1± 0.6 mg/dL to 1.0 ± 0.5 mg/dL [89]. Approximately 42% and 35% of subjects on Pegloticase every two weeks and every four weeks, respectively, achieved the primary efficacy end point of serum urate control [90].

Use of detoxifying enzymes may also hold promise for treating acute life-threatening alcohol intoxication. An enzyme complex containing both alcohol oxidase and catalase was recently reported to reduce blood alcohol levels in intoxicated mice [91]. Alcohol oxidase catalyzes the reaction converting a primary alcohol and oxygen into an aldehyde and hydrogen peroxide. Catalase can then catalyze the decomposition of the toxic intermediate, hydrogen peroxide, into harmless oxygen and water. When given as an antidote, the mice treated with the dually loaded NCs presented with the lowest alanine aminotransferase level and the most significant blood alcohol concentration reduction of 15.8%, 26.1%, 34.7%, and 21.3% at 45 minutes, 90 minutes, 3 hours, and 5 hours after the injection. This was an improvement over both NCs loaded with the individual drugs and delivered simultaneously and dually loaded liposomes.

These examples, spanning the spectrum of development—one that is in clinical use and another in the initial design phase—illustrate the potential of using NCs for this class of enzymes.

5.3.5 Enzyme Replacement Therapy

Numerous diseases, in particular lysosomal storage diseases, arise from the absence of a particular enzyme. While replacing the enzyme seems like the obvious therapy (hence the term "enzyme replacement therapy"), suboptimal delivery has again proven to be a limitation. From the drug delivery standpoint, ERTs can be discussed as interventions whose desirable site of action is located in blood (i.e., systemic) or in the cellular compartments, specifically in the lysosomes.

5.3.5.1 Systemic ERT

Enzyme PEGylation is used for prolongation of ERT circulation, needed for systemic effects. As the best example, pegademase bovine (PEG–adenosine deaminase deficiency [ADA]) is a clinically approved therapy for the treatment of ADA. This rare, inherited disorder of purine metabolism allows accumulation of toxic metabolites that cause defects in the immune system. It is characterized by severe combined immunodeficiency, a failure to thrive, and metabolic abnormalities. PEG-ADA was first introduced as a treatment in 1986 and has been given to over 150 patients worldwide [92, 93]. With a half-life varying from three to six days, the current dosage regimen calls for intramuscular injection once to twice weekly. Although not curative, the level of function achieved is normally sufficient to protect against opportunistic and life-threatening infections. This therapy is particularly valuable for patients who lack a human leukocyte antigen (HLA)-identical bone marrow donor and are judged to be a too high risk for HLA-haploidentical marrow transplantation.

PEGylated phenylalanine ammonia lyase (PEG-PAL) has been developed as a potential therapy for the treatment of phenylketonuria (PKU). It is designed to replace phenylalanine hydroxylase, the enzyme that catalyzes the irreversible conversion of phenylalanine to tyrosine. Without this enzyme, phenylalanine reaches neurotoxic levels and may impair cognitive development. While following a restricted diet may minimize symptoms, this approach suffers from limited compliance. PEG-PAL was shown to reduce blood phenylalanine levels in PKU mouse models [94, 95]. Interestingly, the species with the highest specific activity was not the most effective; stability instead appeared to be the critical factor [94]. BioMarin Pharmaceutical is currently conducting phase 2 clinical trials on this therapy [96–98].

Polymeric NCs have been developed for other enzymes as well. Colonna et al. reported on the use of chitosan glutamate nanoparticles for delivery of prolidase [99–101]. Both prolidase and PEG-prolidase could be loaded in the particles with encapsulation efficiencies of 40–45%. Release was slower with native prolidase, but both showed sustained release over 48 hours. However, while only 67% of the prolidase activity could be restored following a five-day incubation of PD cells with prolidase-loaded particles, 83% could be restored with PEGylated prolidase-loaded particles after only a two-day incubation.

An interesting NC constructed from the amphiphilic triblock copolymer poly(2-methyloxazoline)-block-poly(dimethylsiloxane)-block-poly(2-methyloxazoline) (PMOXA-PDMS-PMOXA), termed "nanoreactor," has been developed for delivery of thymidine phosphorylase [102]. The absence of this enzyme leads to systemic accumulation of the substrates thymidine and deoxyuridine and development of a disease known as mitochondrial neurogastrointestinal encephalomyopathy (MNGIE). By integrating bacterial channel proteins in the polymeric wall, specifically, the nucleoside-specific porin Tsx, the nanoparticles become permeable for substrates and products. The Tsx porin allows efficient transport of the substrates (thymidine and deoxyuridine) into the particle, and the products (thymine and uracil) back out. The particles appeared stable at 37°C in serum for four days with a loss of only

~50% of the enzyme activity. No toxicity was observed in primary hepatocytes at concentrations up to 100 μg polymer/mL, and no inflammatory reaction was observed upon intraperitoneal injection of the nanoreactors in the mice. However, the therapeutic efficacy of this treatment remains to be seen.

5.3.5.2 Lysosomal storage disorders

Lysosomal storage disorders (LSDs) represent a specific subset of genetic deficiency of enzymes localized in the lysosomes, leading to pathological accumulation of their substrates or lack of metabolites. Recombinant ERTs have been developed for many LSDs, but in contrast with systemic applications considered above, these biotherapeutics must be delivered intracellularly. Unfortunately, cellular uptake of recombinant ERTs is not effective.

As with other biotherapeutics discussed above, PEGylated ERTs have been devised. For example, PEG-aspartoacylase has been considered for the treatment of Canavan disease (CD), a neurodegenerative disorder characterized by loss of motor skills, loss of muscle control, and megalocephaly. Following administration of PEG-aspartoacylase to CD mice, statistically significant increases in brain enzyme activity levels and decreases in substrate levels were observed [103]. However, the therapeutic efficacy of the treatment has not yet been reported. PEG-glucocerebrosidase was developed for the treatment of Gaucher disease, an LSD where glucocerebroside is not adequately degraded. The National Institute of Mental Health sponsored a phase 1 clinical trial for the compound [104], but further studies were suspended when Enzon Inc. stopped manufacturing the product.

Liposomes have been adapted for ERT, though not as extensively as for other enzyme classes. Thekkedath et al. reported on the development of glucerocerebrosidase-loaded liposomes for treating Gaucher disease, as described above [105]. The particles were modified with an octadecyl derivative of rhodamine B, a compound known to enhance lysosomal uptake. Microscopy and flow cytometry studies confirmed lysosomal localization of the liposomes and enzyme, respectively, but therapeutic studies have not yet been reported. Similarly, prolidase, an enzyme involved in collagen catabolism, was incorporated into liposomes for potential treatment of prolidase deficiency (PD) [106]. Studies with cultured fibroblasts from PD patients confirmed that all active prolidase encapsulated in the liposomes was taken up by fibroblasts after six days of incubation. No toxicity was observed with cultured skin fibroblasts, but in vivo studies do not appear to have been pursued.

A series of studies has been published on the use of enzyme-loaded NCs targeted to ICAM for treatment of LSDs. ICAM is a leukocyte anchor expressed primarily on the luminal surface of endothelial cells. Upregulated and functionally involved in many disease conditions including inflammation, oxidative stress, atherosclerosis, thrombosis, and metabolic disorders, it potentially represents an attractive target for delivery of drugs designed to treat these pathologies [107]. Interestingly, endothelial cells do not internalize ICAM-1 antibodies, but they do internalize multivalent anti-ICAM conjugates and anti-ICAM-coated NCs through an induction of ICAM-1 clustering and endocytosis [107]. While most of these studies used a simple model NC, polystyrene beads, in which the enzyme is simply adsorbed to the particle, they demonstrated the efficacy of multivalent targeting of ICAM to achieve efficient lysosomal uptake of the enzymes.

This method was initially evaluated for delivering acid sphingomyelinase (ASM) to lysosomes for the treatment of Niemann Pick disease (NPD), a pathology that occurs upon buildup of sphingomyelin in cells. In ASM-deficient cells, anti-ICAM/ASM NCs reduced fluorescent sphingomyelin to background levels, whereas free ASM enabled only partial reduction [108]. In a second study focused on optimizing the size and geometry of NCs, it was found that while disk-shaped particles had longer half-lives in circulation and higher endothelial targeting specificity, spheres were endocytosed more rapidly—a requirement for lysosomal delivery [109]. Further, while micron-sized carriers of either geometry exhibited prolonged residency in prelysosomal compartments, submicron carriers were quickly trafficked to lysosomes. Whereas 1 μm carriers with ASM showed <10% degradation of sphingomyelin two hours after internalization, ASM loaded in 0.1 μm carriers degraded nearly half of the sphingomyelinase, returning levels of the lipid back to baseline. On the basis of these positive results, PLGA nanoparticles were developed as a more clinically translatable alternative. In an ASM knockout mouse model, anti-ICAM/ASM PLGA NCs accumulated in the relevant organs (spleen, liver, and lung are the main targets for type B NPD) within 30 minutes to a significantly greater extent than free ASM [110]. The organ-to-blood localization ratio for ASM in NCs was 12.4±4.6 in the liver, 10.5±1.5 in the spleen, and 59.9±4.2 in the lung. In contrast, for untargeted ASM, the localization ratio was 0.9±0.2 in the liver, 0.5±0.07 in the spleen, and 0.3±0.03 in the lung. Subsequent studies showed that using nanoparticles targeted to both ICAM and transferrin receptors, ASM delivery was enhanced as compared to free enzyme. Interestingly, the overall biodistribution patterns were different from those of single targeted NCs [1].

The anti-ICAM NC strategy was adapted for delivering α-galactosidase (α-Gal) for Fabry disease, a pathology associated with accumulation of blood group B substances (galabiosylceramide and globotriaosylceramide [Gb3]) in body fluids and tissue lysosomes. This results in cerebrovascular, cardiac, and renal impairment, along with endothelial dysfunction. When α-Gal was delivered to mice via anti-ICAM NCs, the liver and lungs received 18.4±1.2%ID and 36.2±3.5%ID, respectively, versus 6.7±0.6%ID and 0.7±0.1%ID when free enzyme was given [111]. In a Fabry disease cell model, a majority of anti-ICAM/α-Gal NCs were confirmed to traffic to endothelial lysosomes (73.5±2.1%) within three hours. Further, the degradation of Gb3 was shown to be enhanced in cells treated with anti-ICAM/ASM NCs (68.9±5.0% reduction) over treatment with untargeted α-Gal (32.4±1.5% reduction) [111].

As a final example, the anti-ICAM NC strategy has been employed to deliver α-glucosidase (GAA) for the treatment of Pompe disease. GAA deficiency leads to aberrant glycogen storage in lysosomes, resulting primarily in liver and striated (cardiac and skeletal) muscle dysfunction. Targeting specificity was confirmed in a Pompe disease cell model, and glycogen degradation was confirmed to be significantly enhanced with anti-ICAM/GAA NCs over nontargeted GAA (73.9±2.3% vs. 26.0±0.9% glycogen reduction, respectively) [112]. Accumulation of GAA in mouse organs was also significantly higher when delivered via an anti-ICAM NC than in the nontargeted form [112]. For example, in cardiac and skeletal muscles, anti-ICAM/GAA NCs exhibited localization ratios of 1.91 ± 0.39, 0.31 ± 0.05, and 0.25 ± 0.03 for the heart, the gastrocnemius muscle, and the quadriceps muscle, respectively, versus nontargeted GAA localization ratios of 0.17 ± 0.01, 0.04 ± 0.01, and 0.04 ± 0.003. The anti-ICAM NCs also increased GAA delivery in other organs: 584-fold in lungs, 67-fold in the spleen, 23-fold in the liver, 10-fold in the heart, and 6-7-fold in the brain.

Similar to other enzyme classes, PEGylation has led to clinical improvements in ERTs. In comparison, incorporation into nanoparticles has received relatively little attention, although it has potential to improve these therapies. In particular, LSDs may benefit specifically from the use of targeted NCs. By using targeting agents that direct the NCs to the appropriate subcellular location, significantly higher amounts of the administered enzyme may reach the site of action.

5.4 Conclusion: Translational Outlook, Challenges, and Perspectives

NCs may find medical utility to improve the PK, optimize the localization, and control the timing of enzymatic drugs. Yet, the successful development of NC-based therapeutics is contingent upon demonstration of strong advantages that outweigh the downsides in clinically relevant animal models and clinical studies. While NCs may provide many desirable drug delivery features (targeting, intracellular delivery, local activation), their industrial and clinical development for enzyme therapies is complicated by many factors. Carriers may cause adverse effects including harmful tissue deposition (e.g., in kidneys, RES, or the microvasculature), unintended activation of host defense (complement, immune response), and damage to target cells. Acceptable in oncological diseases, these issues may constitute too high a risk in other conditions to allow NCs to be used therapeutically. This is particularly true in cases where chronic repetitive treatment may be necessary.

The complexity of NCs also represents a practical challenge. Industrial production may be difficult, and product homogeneity may be a major concern. Complex NCs may also run into additional regulatory complications and costs. These considerations should motivate a critical analysis of the problem in order to identify scenarios with the highest probability of success.

From a practical standpoint, relatively simple and well-characterized NCs (first of all, liposomes) may provide a shortcut to resolve these issues. On the other hand, the material, bioengineering, and pharmaceutical sciences are currently experiencing an exponential rate of progress in the design of a new generation of carriers, some of which may have features unparalleled by the previous formulations. It is tempting to speculate that many enzymatic cargos eventually will be provided with carriers optimally serving their medical goals.

References

1. Treuel, L., Jiang, X., and Nienhaus, G. U. (2013) New views on cellular uptake and trafficking of manufactured nanoparticles, *J. R. Soc. Interface*, **10**.
2. Kostarelos, K., Lacerda, L., Pastorin, G., Wu, W., Wieckowski, S., Luangsivilay, J., Godefroy, S., Pantarotto, D., Briand, J. P., Muller, S., Prato, M., and Bianco, A. (2007) Cellular uptake of functionalized carbon nanotubes is independent of functional group and cell type, *Nat. Nano*, **2**, 108–113.
3. Leroueil, P. R., Berry, S. A., Duthie, K., Han, G., Rotello, V. M., McNerny, D. Q., Baker, J. R. Jr, Orr, B. G., and Holl, M. M. (2008) Wide varieties of cationic nanoparticles induce defects in supported lipid bilayers, *Nano Lett.*, **8**, 420–424.
4. Hobbs, S. K., Monsky, W. L., Yuan, F., Roberts, W. G., Griffith, L., Torchilin, V. P., and Jain, R. K (1998) Regulation of transport pathways in tumor vessels: Role of tumor type and microenvironment, *Proc. Natl. Acad. Sci. U. S. A.*, **95**, 4607–4612.
5. Matsumura, Y., and Maeda, H. (1986) A new concept for macromolecular therapeutics in cancer chemotherapy: mechanism of tumoritropic accumulation of proteins and the antitumor agent SMANCs, *Cancer Res.*, **46**, 6387–6392.
6. Cheng, Z., Al Zaki, A., Hui, J. Z., Muzykantov, V. R., and Tsourkas, A. (2012) Multifunctional nanoparticles: cost versus benefit of adding targeting and imaging capabilities, *Science*, **338**, 903–910.
7. Ryan, S. M., Mantovani, G., Wang, X., Haddleton, D. M., and Brayden, D. J. (2008) Advances in PEGylation of important biotech molecules: delivery aspects, *Exp. Opin. Drug Delivery*, **5**, 371–383.
8. Howard, M. D., Mantovani, G., Wang, X., Haddleton, D. M., and Brayden, D. J. (2008) PEGylation of nanocarrier drug delivery systems: state of the art, *J. Biomed. Nanotechnol.*, **4**, 133–148.

9. Barenholz, Y. (2012) Doxil®: the first FDA-approved nano-drug; lessons learned, *J. Controlled Release*, **160**, 117–134.
10. Gabizon, A., Shmeeda, H., and Barenholz, Y. (2003) Pharmacokinetics of pegylated liposomal doxorubicin: review of animal and human studies, *Clin. Pharmacokinet.*, **42**, 419–436.
11. Hamilton, A. J., Huang, S. L., Warnick, D., Rabbat, M., Kane, B., Nagaraj, A., Klegerman, M., and McPherson, D. D. (2004) Intravascular ultrasound molecular imaging of atheroma components in vivo, *J. Am. Coll. Cardiol.*, **43**, 453–460.
12. Herbst, S. M., Klegerman, M. E., Kim, H., Qi, J., Shelat, H., Wassler, M., Moody, M. R., Yang, C. M., Ge, X., Zou, Y., Kopechek, J. A., Clubb, F. J., Kraemer, D. C., Huang, S., Holland, C. K., McPherson, D. D., and Geng, Y. J. (2009) Delivery of stem cells to porcine arterial wall with echogenic liposomes conjugated to antibodies against cd34 and intercellular adhesion molecule-1, *Mol. Pharm.*, **7**, 3–11.
13. Hughes, M. S., Marsh, J. N., Hall, C. S., Fuhrhop, R. W., Lacy, E. K., Lanza, G. M., and Wickline, S. A. (2005) Acoustic characterization in whole blood and plasma of site-targeted nanoparticle ultrasound contrast agent for molecular imaging, *J. Acoust. Soc. Am.*, **117**, 964–972.
14. Chorny, M., Fishbein, I., Yellen, B. B., Alferiev, I. S., Bakay, M., Ganta, S., Adamo, R., Amiji, M., Friedman, G., and Levy, R. J. (2010) Targeting stents with local delivery of paclitaxel-loaded magnetic nanoparticles using uniform fields, *Proc. Natl. Acad. Sci. U. S. A.*, **107**, 8346–8351.
15. Ma, Y.-H., Wu, S. Y., Wu, T., Chang, Y. J., Hua, M. Y., and Chen, J. P. (2009) Magnetically targeted thrombolysis with recombinant tissue plasminogen activator bound to polyacrylic acid-coated nanoparticles, *Biomaterials*, **30**, 3343–3351.
16. Bayele, H. K., Sakthivel, T., O'Donell, M., Pasi, K. J., Wilderspin, A. F., Lee, C. A., Toth, I., and Florence, A. T. (2005) Versatile peptide dendrimers for nucleic acid delivery, *J. Pharm. Sci.*, **94**, 446–457.
17. Kobayashi, H., and Brechbiel, M. W. (2005) Nano-sized MRI contrast agents with dendrimer cores, *Adv. Drug Delivery Rev.*, **57**, 2271–2286.
18. Lee, C. C., MacKay, J. A., Fréchet, J. M., and Szoka, F. C. (2005) Designing dendrimers for biological applications, *Nat. Biotechnol.*, **23**, 1517–1526.
19. Kataoka, K., Harada, A., and Nagasaki, Y. (2001) Block copolymer micelles for drug delivery: design, characterization and biological significance, *Adv. Drug Delivery Rev.*, **47**, 113–131.

20. Cai, S., Vijayan, K., Cheng, D., Lima, E. M., and Discher, D. E. (2007) Micelles of different morphologies: advantages of worm-like filomicelles of PEO-PCL in paclitaxel delivery, *Pharm. Res.*, **24**, 2099–2109.

21. Simone, E. A., Dziubla, T. D., Colon-Gonzalez, F., Discher, D. E., and Muzykantov, V. R. (2007) Effect of polymer amphiphilicity on loading of a therapeutic enzyme into protective filamentous and spherical polymer nanocarriers, *Biomacromolecules*, **8**, 3914–3921.

22. Geng, Y., Dalhaimer, P., Cai, S., Tsai, R., Tewari, M., Minko, T., and Discher, D. E. (2007) Shape effects of filaments versus spherical particles in flow and drug delivery, *Nat. Nano*, **2**, 249–255.

23. Champion, J. A., and Mitragotri, S. (2006) Role of target geometry in phagocytosis, *Proc. Natl. Acad. Sci. U. S. A.*, **103**, 4930–4934.

24. Berger, H. J., and Pizzo, S. (1988) Preparation of polyethylene glycol-tissue plasminogen activator adducts that retain functional activity: characteristics and behavior in three animal species, *Blood*, **71**, 1641–1647.

25. Rajagopalan, S., Gonias, S. L., and Pizzo, S. V. (1985) A nonantigenic covalent streptokinase-polyethylene glycol complex with plasminogen activator function, *J. Clin. Invest.*, **75**, 413–419.

26. Kajihara, J., Shibata, K., Nakano, Y., Nishimuro, S., and Kato, K. (1994) Physicochemical characterization of PEG-PPG conjugated human urokinase, *Biochim. Biophys. Acta*, **1199**, 202–208.

27. Sakuragawa, N., Shimizu, K., Kondo, K., Kondo, S., and Niwa, M. (1986) Studies on the effect of PEG-modified urokinase on coagulation-fibrinolysis using beagles, *Thromb. Res.*, **41**, 627–635.

28. Moreadith, R. W., and Collen, D. (2003) Clinical development of PEGylated recombinant staphylokinase (PEG–Sak) for bolus thrombolytic treatment of patients with acute myocardial infarction, *Adv. Drug Delivery Rev.*, **55**, 1337–1345.

29. Collen, D., Sinnaeve, P., Demarsin, E., Moreau, H., De Maeyer, M., Jespers, L., Laroche, Y., and Van de Werf, F. (2000) Polyethylene glycol–derivatized cysteine-substitution variants of recombinant staphylokinase for single-bolus treatment of acute myocardial infarction, *Circulation*, **102**, 1766–1772.

30. Vanwetswinkel, S., Plaisance, S., Zhi-Yong, Z., Vanlinthout, I., Brepoels, K., Lasters, I., Collen, D., and Jespers, L. (2000) Pharmacokinetic and thrombolytic properties of cysteine-linked polyethylene glycol derivatives of staphylokinase, *Blood*, **95**, 936–942.

31. Nguyen, P. D., O'Rear, E. A., Johnson, A. E., Lu, R., and Fung, B. M. (1989) Thrombolysis using liposomal-encapsulated streptokinase: an in vitro study, *Proc. Soc. Exp. Biol. Med.*, **192**, 261–269.

32. Perkins, W. R., Vaughan, D. E., Plavin, S. R., Daley, W. L., Rauch, J., Lee, L., and Janoff, A. S. (1997) Streptokinase entrapment in interdigitation-fusion liposomes improves thrombolysis in an experimental rabbit model, *Thromb. Haemost.*, **77**, 1174–1178.

33. Kim, I.-S., Choi, H. G., Choi, H. S., Kim, B. K., and Kim, C. K. (1998) Prolonged systemic delivery of streptokinase using liposome, *Arch. Pharm. Res.*, **21**, 248–252.

34. Kim, J.-Y., Kim, J. K., Park, J. S., Byun, Y., and Kim, C. K. (2009) The use of PEGylated liposomes to prolong circulation lifetimes of tissue plasminogen activator, *Biomaterials*, **30**, 5751–5756.

35. Heeremans, J. L., Prevost, R., Bekkers, M. E., Los, P., Emeis, J. J., Kluft, C., and Crommelin, D. J. (1995) Thrombolytic treatment with tissue-type plasminogen activator (t-PA) containing liposomes in rabbits: a comparison with free t-PA, *Thromb. Haemost.*, **73**, 488–494.

36. Leach, J. K., O'Rear, E. A., Patterson, E., Miao, Y., and Johnson, A. E. (2003) Accelerated thrombolysis in a rabbit model of carotid artery thrombosis with liposome-encapsulated and microencapsulated streptokinase, *Thromb. Haemost.*, **90**, 64–70.

37. Wang, S.-S., Chou, N.-K., and Chung, T.-W. (2009) The t-PA-encapsulated PLGA nanoparticles shelled with CS or CS-GRGD alter both permeation through and dissolving patterns of blood clots compared with t-PA solution: an in vitro thrombolysis study, *J. Biomed. Mater. Res. A*, **91A**, 753–761.

38. Jin, H., Tan, H., Zhao, L., Sun, W., Zhu, L., Sun, Y., Hao, H., Xing, H., Liu, L., Qu, X., Huang, Y., and Yang, Z. (2012) Ultrasound-triggered thrombolysis using urokinase-loaded nanogels, *Int. J. Pharm.*, **434**, 384–390.

39. Fernandes, E. G., de Queiroz, A. A., Abraham, G. A., and San Román, J. (2006) Antithrombogenic properties of bioconjugate streptokinase-polyglycerol dendrimers, *J. Mater. Sci. Mater. Med.*, **17**, 105–111.

40. Wang, X., Inapagolla, R., Kannan, S., Lieh-Lai, M., and Kannan, R. M. (2007) Synthesis, characterization, and in vitro activity of dendrimer–streptokinase conjugates, *Bioconjugate Chem.*, **18**, 791–799.

41. Huang, G., Zhou, Z., Srinivasan, R., Penn, M. S., Kottke-Marchant, K., Marchant, R. E., and Gupta, A. S. (2008) Affinity manipulation of surface-conjugated RGD peptide to modulate binding of liposomes to activated platelets, *Biomaterials*, **29**, 1676–1685.

42. Vaidya, B., Agrawal, G. P., and Vyas, S. P. (2011) Platelets directed liposomes for the delivery of streptokinase: development and characterization, *Eur. J. Pharm. Sci.*, **44**, 589–594.

43. Yan, H.-L., Sun, S. H., Wang, W. T., Ding, F. X., Mei, Q., and Xue, G. (2007) Design and characterization of a platelet-targeted plasminogen activator with enhanced thrombolytic and antithrombotic potency, *Biotechnol. Appl. Biochem.*, **46**, 115–125.

44. Zhu, J., Xue, J., Guo, Z., Zhang, L., and Marchant, R. E. (2007) Biomimetic glycoliposomes as nanocarriers for targeting p-selectin on activated platelets, *Bioconjugate Chem.*, **18**, 1366–1369.

45. Marsh, J. N., Hu, G., Scott, M. J., Zhang, H., Goette, M. J., Gaffney, P. J., Caruthers, S. D., Wickline, S. A., Abendschein, D., and Lanza, G. M. (2011) A fibrin-specific thrombolytic nanomedicine approach to acute ischemic stroke, *Nanomedicine*, **6**, 605–615.

46. McCarthy, J. R., Sazonova, I. Y., Erdem, S. S., Hara, T., Thompson, B. D., Patel, P., Botnaru, I., Lin, C. P., Reed, G. L., Weissleder, R., and Jaffer, F. A. (2012) Multifunctional nanoagent for thrombus-targeted fibrinolytic therapy, *Nanomedicine*, **7**, 1017–1028.

47. Laing, S. T., Moody, M. R., Kim, H., Smulevitz, B., Huang, S. L., Holland, C. K., McPherson, D. D., and Klegerman, M. E. (2012) Thrombolytic efficacy of tissue plasminogen activator-loaded echogenic liposomes in a rabbit thrombus model, *Thromb. Res.*, **130**, 629–635.

48. Shaw, G. J., Meunier, J. M., Huang, S. L., Lindsell, C. J., McPherson, D. D., and Holland, C. K. (2009) Ultrasound-enhanced thrombolysis with tPA-loaded echogenic liposomes, *Thromb. Res.*, **124**, 306–310.

49. Tiukinhoy-Laing, S. D., Huang, S., Klegerman, M., Holland, C. K., and McPherson, D. D. (2007) Ultrasound-facilitated thrombolysis using tissue-plasminogen activator-loaded echogenic liposomes, *Thromb. Res.*, **119**, 777–784.

50. Uesugi, Y., Kawata, H., Jo, J., Saito, Y., and Tabata, Y. (2010) An ultrasound-responsive nano delivery system of tissue-type plasminogen activator for thrombolytic therapy, *J. Controlled Release*, **147**, 269–277.

51. Härdig, B. M., Persson, H. W., and Olsson, S. B. (2006) Low-energy ultrasound exposure of the streptokinase molecule may enhance but also attenuate its fibrinolytic properties, *Thromb. Res.*, **117**, 713–720.

52. Nilsson, A. M., Odselius, R., Roijer, A., and Olsson, S. B. (1995) Pro- and antifibrinolytic effects of ultrasound on streptokinase-induced thrombolysis, *Ultrasound Med. Biol.*, **21**, 833–840.

53. Bi, F., Zhang, J., Su, Y., Tang, Y. C., and Liu, J. N. (2009) Chemical conjugation of urokinase to magnetic nanoparticles for targeted thrombolysis, *Biomaterials*, **30**, 5125–5130.

54. Absar, S., Choi, S., Yang, V. C., and Kwon, Y. M. (2012) Heparin-triggered release of camouflaged tissue plasminogen activator for targeted thrombolysis, *J. Controlled Release*, **157**, 46–54.

55. Korin, N., Kanapathipillai, M., Matthews, B. D., Crescente, M., Brill, A., Mammoto, T., Ghosh, K., Jurek, S., Bencherif, S. A., Bhatta, D., Coskun, A. U., Feldman, C. L., Wagner, D. D., and Ingber, D. E. (2012) Shear-activated nanotherapeutics for drug targeting to obstructed blood vessels, *Science*, **337**, 738–742.

56. Holme, M. N., Fedotenko, I. A., Abegg, D., Althaus, J., Babel, L., Favarger, F., Reiter, R., Tanasescu, R., Zaffalon, P. L., Ziegler, A., Müller, B., Saxer, T., and Zumbuehl, A. (2012) Shear-stress sensitive lenticular vesicles for targeted drug delivery, *Nat. Nano*, **7**, 536–543.

57. Muzykantov, V. R. (2001) Targeting of superoxide dismutase and catalase to vascular endothelium, *J. Controlled Release*, **71**, 1–21.

58. Freeman, B. A., Young, S. L., and Crapo, J. D. (1983) Liposome-mediated augmentation of superoxide dismutase in endothelial cells prevents oxygen injury, *J. Biol. Chem.*, **258**, 12534–12542.

59. Laursen, J. B., Rajagopalan, S., Galis, Z., Tarpey, M., Freeman, B. A., and Harrison, D. G. (1997) Role of superoxide in angiotensin II-induced but not catecholamine-induced hypertension, *Circulation*, **95**, 588–593.

60. Corvo, L. M., Jorge, J. C., van't Hof, R., Cruz, M. E., Crommelin, D. J., and Storm, G. (2002) Superoxide dismutase entrapped in long-circulating liposomes: formulation design and therapeutic activity in rat adjuvant arthritis, *Biochim. Biophys. Acta.*, **1564**, 227–236.

61. Gaspar, M. M., Martins, M. B., Corvo, M. L., and Cruz, M. E. (2003) Design and characterization of enzymosomes with surface-exposed superoxide dismutase, *Biochim. Biophys. Acta*, **1609**, 211–217.

62. Gaspar, M. M., Boerman, O. C., Laverman, P., Corvo, M. L., Storm, G., and Cruz, M. E. (2007) Enzymosomes with surface-exposed superoxide dismutase: in vivo behaviour and therapeutic activity in a model of adjuvant arthritis, *J. Controlled Release*, **117**, 186–195.

63. Xu, X., Costa, A., and Burgess, D. J. (2012) Protein encapsulation in unilamellar liposomes: high encapsulation efficiency and a novel technique to assess lipid-protein interaction, *Pharm. Res.*, **29**, 1919–31.

64. Batrakova, E. V., Li, S., Reynolds, A. D., Mosley, R. L., Bronich, T. K., Kabanov, A. V., and Gendelman, H. E. (2007) A macrophage-nanozyme delivery system for Parkinson's disease, *Bioconjugate Chem.*, **18**, 1498–1506.

65. Brynskikh, A. M., Zhao, Y., Mosley, R. L., Li, S., Boska, M. D., Klyachko, N. L., Kabanov, A. V., Gendelman, H. E., and Batrakova, E. V. (2010) Macrophage delivery of therapeutic nanozymes in a murine model of Parkinson's disease, *Nanomedicine*, **5**, 379–396.

66. Rosenbaugh, E. G., Roat, J. W., Gao, L., Yang, R. F., Manickam, D. S., Yin, J. X., Schultz, H. D., Bronich, T. K., Batrakova, E. V., Kabanov, A. V., Zucker, I. H., and Zimmerman, M. C. (2010) The attenuation of central angiotensin II-dependent pressor response and intra-neuronal signaling by intracarotid injection of nanoformulated copper/zinc superoxide dismutase, *Biomaterials*, **31**, 5218–5226.

67. Batrakova, E. V., Li, S., Miller, D. W., and Kabanov, A. V. (1999) Pluronic P85 increases permeability of a broad spectrum of drugs in polarized BBMEC and Caco-2 cell monolayers, *Pharm. Res.*, **16**, 1366–1372.

68. Batrakova, E. V., Miller, D. W., Li, S., Alakhov, V. Y., Kabanov, A. V., and Elmquist, W. F. (2001) Pluronic P85 enhances the delivery of digoxin to the brain: in vitro and in vivo studies, *J. Pharmacol. Exp. Ther.*, **296**, 551–557.

69. Batrakova, E. V., Zhang, Y., Li, Y., Li, S., Vinogradov, S. V., Persidsky, Y., Alakhov, V. Y., Miller, D. W., and Kabanov, A. V. (2004) Effects of pluronic P85 on GLUT1 and MCT1 transporters in the blood-brain barrier, *Pharm. Res.*, **21**, 1993–2000.

70. Yi, X., Zimmerman, M. C., Yang, R., Tong, J., Vinogradov, S., and Kabanov, A. V. (2010) Pluronic-modified superoxide dismutase 1 attenuates angiotensin II-induced increase in intracellular superoxide in neurons, *Free Rad. Biol. Med.*, **49**, 548–558.

71. Yan, M., Du, J., Gu, Z., Liang, M., Hu, Y., Zhang, W., Priceman, S., Wu, L., Zhou, Z. H., Liu, Z., Segura, T., Tang, Y., and Lu, Y. (2010) A novel intracellular protein delivery platform based on single-protein nanocapsules, *Nat. Nanotechnol.*, **5**, 48–53.

72. Chorny, M., Hood, E., Levy, R. J., and Muzykantov, V. R. (2010) Endothelial delivery of antioxidant enzymes loaded into non-polymeric magnetic nanoparticles, *J. Controlled Release*, **146**, 144–151.

73. Giovagnoli, S., Luca, G., Casaburi, I., Blasi, P., Macchiarulo, G., Ricci, M., Calvitti, M., Basta, G., Calafiore, R., and Rossi, C. (2005) Long-term delivery of superoxide dismutase and catalase entrapped in poly(lactide-co-glycolide) microspheres: in vitro effects on isolated neonatal porcine pancreatic cell clusters, *J. Controlled Release*, **107**, 65–77.

74. Dziubla, T. D., Karim, A., and V. R. Muzykantov (2005) Polymer nanocarriers protecting active enzyme cargo against proteolysis, *J. Controlled Release*, **102**, 427–439.

75. Dziubla, T. D., Shuvaev, V. V., Hong, N. K., Hawkins, B. J., Madesh, M., Takano, H., Simone, E., Nakada, M. T., Fisher, A., Albelda, S. M., and Muzykantov, V. R. (2008) Endothelial targeting of semi-permeable polymer nanocarriers for enzyme therapies, *Biomaterials*, **29**, 215–227.

76. Avramis, V. I., and Tiwari, P. N. (2006) Asparaginase (native ASNase or pegylated ASNase) in the treatment of acute lymphoblastic leukemia, *Int. J. Nanomed.*, **1**, 241–254.

77. Cruz, M. E. M., Gaspara, M. M., Lopesb, F., Jorgea, J. S.,and Perez-Solerc, R. (1993) Liposomal l-asparaginase: in vitro evaluation, *Int. J. Pharm.*, **96**, 67–77.

78. Gaspar, M. M., Blanco, D., Cruz, M. E., and Alonso, M. J. (1998) Formulation of L-asparaginase-loaded poly(lactide-co-glycolide) nanoparticles: influence of polymer properties on enzyme loading, activity and in vitro release, *J. Controlled Release*, **52**, 53–62.

79. Steinhilber, D., Witting, M., Zhang, X., Staegemann, M., Paulus, F., Friess, W., Küchler, S., and Haag, R. (2012) Surfactant free preparation of biodegradable dendritic polyglycerol nanogels by inverse nanoprecipitation for encapsulation and release of pharmaceutical biomacromolecules, *J. Controlled Release*. **169**(3), 289–295.

80. Glazer, E. S., Piccirillo, M., Albino, V., Di Giacomo, R., Palaia, R., Mastro, A. A., Beneduce, G., Castello, G., De Rosa, V., Petrillo, A., Ascierto, P. A., Curley, S. A., and Izzo, F. (2010) Phase II study of pegylated arginine deiminase for nonresectable and metastatic hepatocellular carcinoma, *J. Clin. Oncol.*, **28**, 2220–2226.

81. Yang, T. S., Lu, S. N., Chao, Y., Sheen, I. S., Lin, C. C., Wang, T. E., Chen, S. C., Wang, J. H., Liao, L. Y., Thomson, J. A., Wang-Peng, J., Chen, P. J., and Chen, L. T. (2010) A randomised phase II study of pegylated arginine deiminase (ADI-PEG 20) in Asian advanced hepatocellular carcinoma patients, *Br. J. Cancer*, **103**, 954–960.

82. Ascierto, P. A., Scala, S., Castello, G., Daponte, A., Simeone, E., Ottaiano, A., Beneduce, G., De Rosa, V., Izzo, F., Melucci, M. T., Ensor, C. M., Prestayko, A. W., Holtsberg, F. W., Bomalaski, J. S., Clark, M. A., Savaraj, N., Feun, L. G., and Logan, T. F. (2005) Pegylated arginine deiminase treatment of patients with metastatic melanoma: results from phase I and II studies, *J. Clin. Oncol.*, **23**, 7660–7668.

83. Ott, P. A., Carvajal, R. D., Pandit-Taskar, N., Jungbluth, A. A., Hoffman, E. W., Wu, B. W., Bomalaski, J. S., Venhaus, R., Pan, L., Old, L. J., Pavlick, A. C., and Wolchok, J. D. (2012) Phase I/II study of pegylated arginine deiminase (ADI-PEG 20) in patients with advanced melanoma, *Invest. New Drugs*, 1–10.

84. Oz, H. S., J. Zhong, and de Villiers, W. J. S. (2012) Pegylated arginine deiminase downregulates colitis in murine models, *Mediators Inflamm.*, **2012**.

85. Izzo, F., Montella, M., Orlando, A. P., Nasti, G., Beneduce, G., Castello, G., Cremona, F., Ensor, C. M., Holtzberg, F. W., Bomalaski, J. S., Clark, M. A., Curley, S. A., Orlando, R., Scordino, F., and Korba, B. E. (2007) Pegylated arginine deiminase lowers hepatitis C viral titers and inhibits nitric oxide synthesis, *J. Gastroenterol. Hepatol.*, **22**, 86–91.

86. Abuchowski, A., Karp, D., and Davis, F. F. (1981) Reduction of plasma urate levels in the cockerel with polyethylene glycol-uricase, *J. Pharmacol. Exp. Ther.*, **219**, 352–354.

87. Nishimura, H., Ashihara, Y., Matsushima, A., and Inada, Y. (1979) Modification of yeast uricase with polyethylene glycol: disappearance of binding ability towards anti-uricase serum, *Enzyme*, **24**, 261–264.

88. Schlesinger, N., Yasothan, U., and Kirkpatrick, P. (2011) Pegloticase, *Nat. Rev. Drug Discovery*, **10**, 17–18.

89. Sundy, J. S., Ganson, N. J., Kelly, S. J., Scarlett, E. L., Rehrig, C. D., Huang, W., and Hershfield, M. S. (2007) Pharmacokinetics and pharmacodynamics of intravenous PEGylated recombinant mammalian urate oxidase in patients with refractory gout, *Arthritis Rheum.*, **56**, 1021–1028.

90. Reinders, M. K., and Jansen, T. L. (2010) New advances in the treatment of gout: review of pegloticase, *Ther. Clin. Risk Manag.*, **2010**, 543–550.

91. Liu, Y., Du, J., Yan, M., Lau, M. Y., Hu, J., Han, H., Yang, O. O., Liang, S., Wei, W., Wang, H., Li, J., Zhu, X., Shi, L., Chen, W., Ji, C., and Lu, Y. (2013) Biomimetic enzyme nanocomplexes and their use as antidotes and preventive measures for alcohol intoxication, *Nat. Nano*, **8**, 187–192.

92. Booth, C., and Gaspar, H. B. (2009) Pegademase bovine (PEG-ADA) for the treatment of infants and children with severe combined immunodeficiency (SCID), *Biologics*, **3**, 349–358.

93. Hershfield, M. S. (1995) PEG-ADA replacement therapy for adenosine deaminase deficiency: an update after 8.5 years, *Clin. Immunol. Immunopathol.*, **76**, S228–S232.

94. Sarkissian, C. N., Gámez, A., Wang, L., Charbonneau, M., Fitzpatrick, P., Lemontt, J. F., Zhao, B., Vellard, M., Bell, S. M., Henschell, C., Lambert, A., Tsuruda, L., Stevens, R. C., and Scriver, C. R. (2008) Preclinical evaluation of multiple species of PEGylated recombinant phenylalanine ammonia lyase for the treatment of phenylketonuria, *Proc. Natl. Acad. Sci. U. S. A.*, **105**, 20894–20899.

95. Wang, L., Gamez, A., Sarkissian, C. N., Straub, M., Patch, M. G., Han, G. W., Striepeke, S., Fitzpatrick, P., Scriver, C. R., and Stevens, R. C. (2005) Structure-based chemical modification strategy for enzyme replacement treatment of phenylketonuria, *Mol. Genet. Metab.*, **86**, 134–140.

96. BioMarin Pharmaceutical (2000) Phase 2, open-label dose-finding study to evaluate the safety, efficacy, and tolerability of multiple subcutaneous (SC) doses of rAvPAL-PEG in subjects with PKU, in ClinicalTrials.gov [Internet]. National Library of Medicine, Bethesda, MD (cited Aug. 29, 2013). Available from http://clinicaltrials.gov/show/NCT00925054. NLM identifier: NCT00925054.

97. BioMarin Pharmaceutical (2000) Safety, tolerability, and efficacy study of rAvPAL-PEG administered daily in subjects with phenylketonuria (PKU), in ClinicalTrials.gov [Internet]. National Library of Medicine, Bethesda, MD (cited Mar. 31, 2013). Available from http://clinicaltrials.gov/show/NCT01212744. NLM identifier: NCT01212744.

98. BioMarin Pharmaceutical (2000) A study to evaluate subcutaneously administered rAvPAL-PEG in patients with phenylketonuria for 24 weeks, in ClinicalTrials.gov [Internet]. National Library of Medicine, Bethesda, MD [cited Aug. 29, 2013). Available from http://clinicaltrials.gov/show/NCT01560286. NLM identifier: NCT01560286.

99. Colonna, C., Conti, B., Perugini, P., Pavanetto, F., Modena, T., Dorati, R., and Genta, I. (2007) Chitosan glutamate nanoparticles for protein delivery: Development and effect on prolidase stability, *J. Microencapsul.*, **24**, 553–564.

100. Colonna, C., Conti, B., Perugini, P., Pavanetto, F., Modena, T., Dorati, R., Iadarola, P., and Genta, I. (2008) Ex vivo evaluation of prolidase loaded chitosan nanoparticles for the enzyme replacement therapy, *Eur. J. Pharm. Biopharm.*, **70**, 58–65.

101. Colonna, C., Conti, B., Perugini, P., Pavanetto, F., Modena, T., Dorati, R., Iadarola, P., and Genta, I. (2008) Site-directed PEGylation as successful approach to improve the enzyme replacement in the case of prolidase, *Int. J. Pharm.*, **358**, 230–237.

102. De Vocht, C., Ranquin, A., Willaert, R., Van Ginderachter, J. A., Vanhaecke, T., Rogiers, V., Versées, W., Van Gelder, P., and Steyaert, J. (2009) Assessment of stability, toxicity and immunogenicity of new polymeric nanoreactors for use in enzyme replacement therapy of MNGIE, *J. Controlled Release*, **137**, 246–254.

103. Zano, S., Malik, R., Szucs, S., Matalon, R., and Viola RE. (2011) Modification of aspartoacylase for potential use in enzyme replacement therapy for the treatment of Canavan disease, *Mol. Genet. Metab.*, **102**, 176–180.

104. National Institute of Mental Health (2000) A phase I and II study of PEG-glucocerebrosidase in patients with type 1 or type 3 Gaucher disease, in ClinicalTrials.gov [Internet]. National Library of Medicine, Bethesda, MD (cited Mar. 3, 2008). Available from http://clinicaltrials.gov/show/ NCT00001410. NLM identifier: NCT00001410.

105. Thekkedath, R., Koshkaryev, A., and Torchilin, V. P. (2012) Lysosome-targeted octadecyl-rhodamine B-liposomes enhance lysosomal accumulation of glucocerebrosidase in Gaucher's cells in vitro, *Nanomedicine.* **8**(7), 1055–1065.

106. Perugini, P., Hassan, K., Genta, I., Modena, T., Pavanetto, F., Cetta, G., Zanone, C., Iadarola, P., Asti, A., and Conti, B. (2005) Intracellular delivery of liposome-encapsulated prolidase in cultured fibroblasts from prolidase-deficient patients, *J. Controlled Release*, **102**, 181–190.

107. Muro, S., Wiewrodt, R., Thomas, A., Koniaris, L., Albelda, S. M., Muzykantov, V. R., and Koval, M. (2003) A novel endocytic pathway induced by clustering endothelial ICAM-1 or PECAM-1, *J. Cell Sci.*, **116**, 1599–1609.

108. Muro, S., Schuchman, E. H., and Muzykantov, V. R. (2006) Lysosomal enzyme delivery by icam-1-targeted nanocarriers bypassing glycosylation- and clathrin-dependent endocytosis, *Mol. Ther.*, **13**, 135–141.

109. Muro, S., Garnacho, C., Champion, J. A., Leferovich, J., Gajewski, C., Schuchman, E. H., Mitragotri, S., and Muzykantov, V. R. (2008) Control of endothelial targeting and intracellular delivery of therapeutic enzymes by modulating the size and shape of icam-1-targeted carriers, *Mol. Ther.*, **16**, 1450–1458.

110. Garnacho, C., Dhami, R., Simone, E., Dziubla, T., Leferovich, J., Schuchman, E. H., Muzykantov, V., and Muro, S. (2008) Delivery of acid sphingomyelinase in normal and niemann-pick disease mice using intercellular adhesion molecule-1-targeted polymer nanocarriers, *J. Pharmacol. Exp. Ther.*, **325**, 400–408.

111. Hsu, J., Serrano, D., Bhowmick, T., Kumar, K., Shen, Y., Kuo, Y. C., Garnacho, C., and Muro, S. (2011) Enhanced endothelial delivery and biochemical effects of α-galactosidase by ICAM-1-targeted nanocarriers for Fabry disease, *J. Controlled Release*, **149**, 323–331.

112. Hsu, J., Northrup, L., Bhowmick, T., and Muro, S. (2012) Enhanced delivery of α-glucosidase for Pompe disease by ICAM-1-targeted nanocarriers: comparative performance of a strategy for three distinct lysosomal storage disorders, *Nanomed.: NBM*, **8**, 731–739.

Chapter 6

DNA-Origamis as Protein Nanocarriers

Juan Elezgaray,[a] Jean-Pierre Aimé,[a] and Jean-Michel Arbona[b]
[a]*CBMN, UMR 5248, CNRS, Allée de St Hilaire, Bat. B14 Bis, F-33600 Pessac, France*
[b]*Insitut Pasteur, 25-28 Rue du Docteur Roux, 75015 Paris, France*
j.elezgaray@iecb.u-bordeaux.fr

DNA-based nanotechnologies are bottom-up approaches that use nucleic acids as building blocks. The high specificity of Watson–Crick interactions is the key element to understand the huge range of diverse nanometric constructions that have been designed and experimentally characterized so far. In this context, the "origami" strategy, initially introduced as a generic method to design almost arbitrary planar shapes, is particularly important [1]. Small oligonucleotides staples recognizing specific sequences drive the folding of a long DNA strand. A useful property of DNA origamis is the possibility to arrange, with nanometric precision, any kind of molecular-sized objects that can be bound in some way to DNA. A nonexhaustive list of these objects includes gold nanoparticles, carbon nanotubes, aptamers, and proteins. It is obvious that this property makes DNA origamis particularly promising as a platform to spatially arrange proteins with high precision and specificity. The

Enzyme Nanocarriers
Edited by Daniela Cardinale and Thierry Michon
Copyright © 2016 Pan Stanford Publishing Pte. Ltd.
ISBN 978-981-4613-42-2 (Hardcover), 978-981-4613-43-9 (eBook)
www.panstanford.com

goal of this chapter is to provide a glimpse of published work in these directions, both in vitro and in vivo, as well as to point out the possible extensions and expected difficulties.

6.1 A Quick Primer to DNA Origami Formation

The fundamental building block of DNA-based nanostructures is the double helix. In the early attempts to introduce diversity in DNA-made structures, flexible structures were engineered [2] that could self-assemble in 3D lattices. More rigid tiles [3] could arrange in planar lattices of programmable periodicity [4, 5] or fold into nanotubes [6]. However, all these contributions required a specific design. No strategy had been yet developed for the assembly of various kinds of patterns, starting from a single raw DNA material. From this point of view, the contribution of Rothemund, with the origami strategy [1], can be seen as a landmark in the development of bioinspired bottom-up methods.

In 2006, Rothemund showed [1] that appropriate sets of helper strands (single-stranded DNA [ssDNA]) are able to fold long ssDNA scaffolds, such as the 7429 nt M13 phage genome. The choice of each set of staples is dictated by the final shape. It follows quite simple rules, which eventually boil down to the requirement of having, locally, double-stranded helices, as in the usual B-DNA conformation. This approach is remarkable for several reasons: (i) the yield of the method is close to 100% in many cases, (ii) it is robust to impurities in the staple composition, and (iii) it is extremely versatile, as the same scaffold can be used to design countless planar structures, with the possibility to act as a pegboard where other molecules can be grafted.

The exact mechanism of origami formation is not completely understood. An annealing step is necessary to first destabilize any secondary structure of the scaffold and then to progressively allow the binding of the staples (Fig. 6.1). The role of staples is to hold together separate parts of the scaffold; the main hindrance to their binding comes from the difficulty of one staple to approach at the same time sequences of bases that due to random motion of the scaffold can be quite far, one from the other. The origami formation proceeds in such a way that more stable staples (GC rich) bind first (at higher temperature). This initial binding helps [7–9] the less stable staples to bind to the scaffold, as its motion becomes more and more constrained as the binding proceeds. Recent experiments show that for each set of staples there exists a temperature for which the annealing can be done very quickly [10] with almost 100% yield.

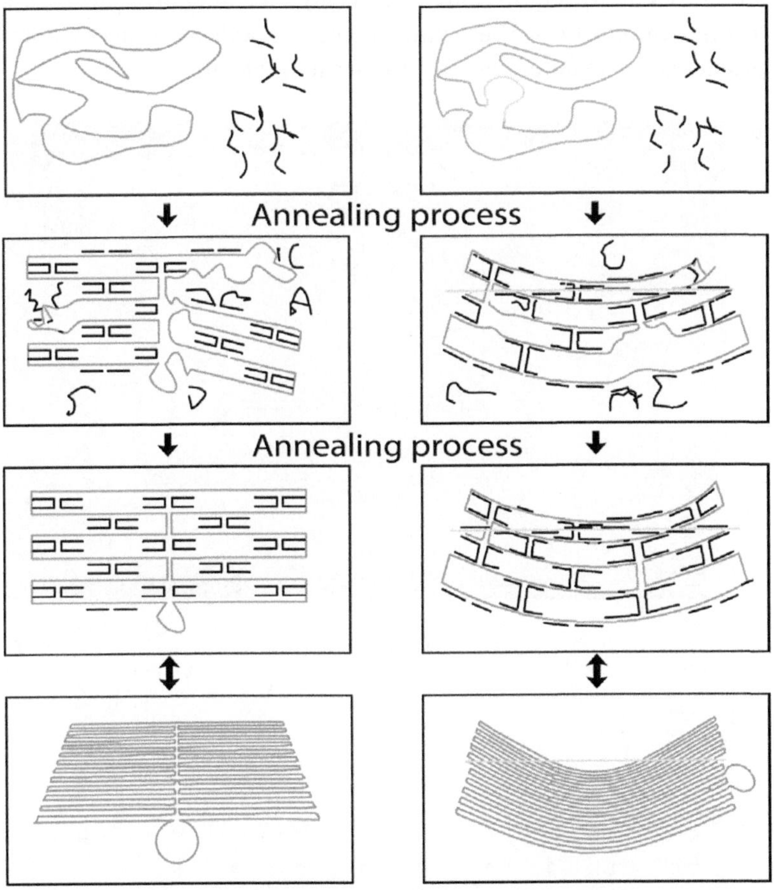

Figure 6.1 A schematic description of the origami-folding process. A set of short helper strands (black) are mixed with a long ssDNA scaffold (red). As the temperature is lowered (follow the arrows), staples successively bind to the scaffold to reach the final shape at the bottom. The second column shows the same process for a rectangle that is bent (by construction) at the end of the annealing process.

6.2 Boxes and Nanorobots

The original idea was quickly extended to obtain 3D nanostructures, ranging from hollow 3D cages [11] to arrangements of helices along predefined curved surfaces [12] or patterns from honeycomb arrangements (Fig. 6.2) [13]. Contrary to the planar shapes, DNA structures in 3D require long (few days) annealing procedures, due to the complexity of the arrangements as well as to accessibility problems as the outer layers of the structure form.

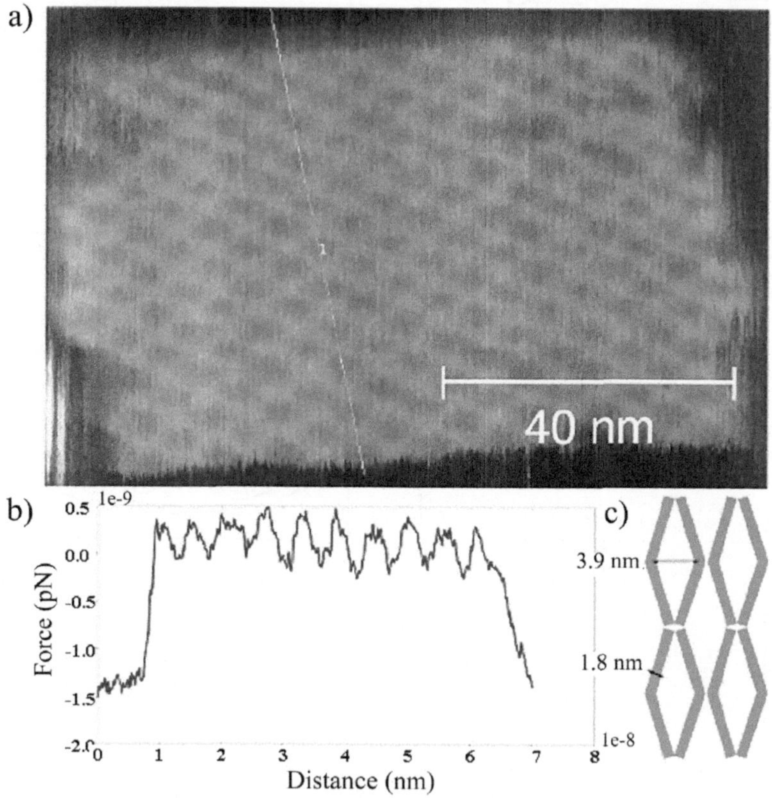

Figure 6.2 (a) An AFM image of a rectangular origami deposited on a mica surface. It shows a typical honeycomb-like pattern, which characterizes the arrangement under tension of the double helices that form the structure. (b) A section of the previous 2D picture along the blue line. (c) Schematic representation of the dsDNA strands as measured by AFM.

The first example of a DNA box [11] gathered many of the desirable properties of a potential drug delivery container. It was designed by folding the M13 scaffold into an empty cube, each of the six faces being obtained much in the way planar origamis were obtained in Ref. [6] with a final size of $42 \times 36 \times 36$ nm^3. One of the faces acted as a lid, which was closed right after annealing but could be opened in the presence of an ssDNA using strand displacement. However, this initial design suffered from at least two difficulties: The box was floppy and leaky, a major inconvenience from the drug delivery point of view; second, the opening of the lid was irreversible. A second attempt [14] led to an improved design, where a smaller box ($18 \times 18 \times 24$ nm^3) was obtained folding a 1983 nt ssDNA, derived from a truncated pUC plasmid. The closing and opening system was also improved, being ensured by two sets of keys (in both cases, these are ssDNA strands), which could displace one to each other through binding to an eight-base overhang. The open–close cycle could be repeated for three rounds.

6.2.1 Nanorobots for Drug Delivery

A similar approach [15] leads to a DNA nanorobot that could deliver payloads after reconfiguring its structure in a triggered and conditional activation (Fig. 6.3). Interestingly, the capabilities of this nanorobot could be tested in vitro, providing a proof of principle for drug delivery, enhanced by logic capabilities. The design of the cage is different from [11] and [14]: a hexagonal barrel with dimensions $35 \times 35 \times 45$ nm^3 was designed using the program cadnano [16]. The two halves are connected by a flexible hinge. The closing is ensured by two aptamer lock mechanisms located at the side opposite the hinge: one of the two halves contains the two aptamers; the other half contains partially complementary strands that ensure the closing of the structure (two additional helpers are actually needed to overcome the entropic elasticity). The initial conformation of the two aptamers can be destabilized in the presence of an antigen key in the 10 pM range. As in the two previous box constructions, the opening is driven by entropic elasticity. In Ref. [15], the authors loaded the structure with 5 nm gold nanoparticles and various Fab' antibody fragments against human leukocyte antigen (HLA)-A/B/C, fluorescently tagged. Three aptamers (41t [17], TE17 [18], and sgc8c [19]) targeting different "key" antigens were used. In the presence of any of these factors, the corresponding aptamer would change of conformation, leading to unzipping from the partial complement. Because the presence of two factors induced a simultaneous folding change of two aptamers and effective opening of the robot, it was effectively driven by an AND logic gate. The opening could be experimentally monitored by fluorescent measurements.

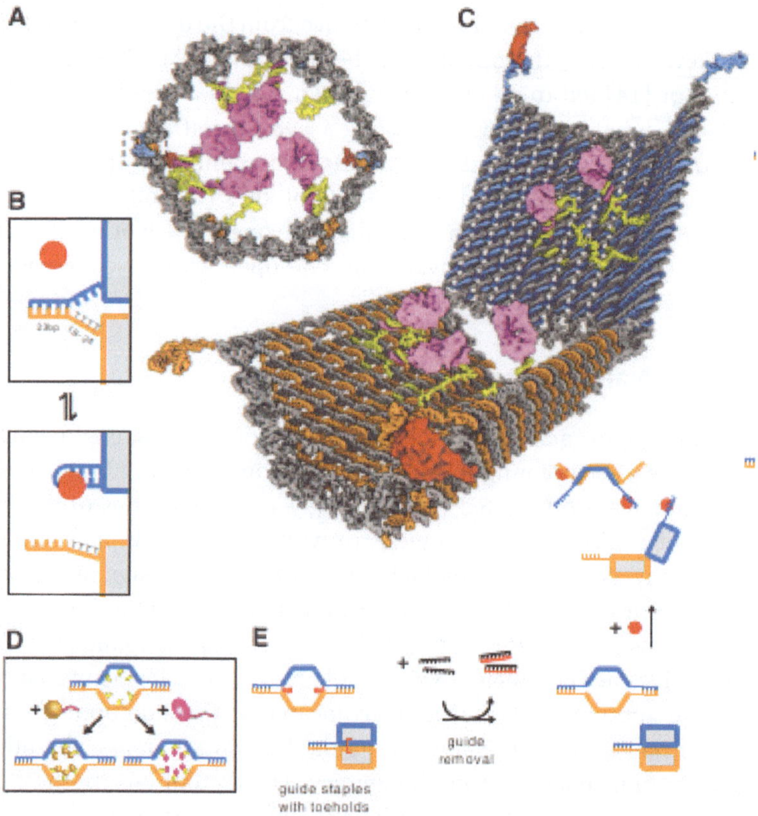

Figure 6.3 The nanorobot designed in Ref. [15]. (A) A transversal section. The scaffold is gray. (B) Aptamers act as locks of the structure. They change conformation upon binding with a ligand (red circle), leading to opening of the nanorobot. (C) Side view of the opened state, showing the cargos (gold nanoparticles and fluorescent antibodies). Reproduced from Ref. [15], Copyright 2009 The Authors.

6.3 Protein and Nucleic Acid Arrays

Many enzymatic cascades that influence cellular activities are strongly influenced by the spatial arrangement of their components. It is believed that this organization facilitates substrate channeling [20], avoiding time-consuming diffusion-limited transport and limiting cross talk between different signaling pathways. There are only a few methods that allow this precise positioning to be done in a reproducible manner. Other chapters in this book account for several of these methods based on the use of virus scaffolds. DNA origamis are also natural candidates for this task. Indeed, DNA origamis can be seen as platforms where attaching points can be positioned with nanometric precision. First attempts to bind biological components have been limited to nucleic acids (DNA or RNA) [21].

6.3.1 Arranging Nucleic Acid Probes

In Ref. [21], the authors focused on the detection of low levels of gene expression, particularly the presence of low concentrations of mRNAs. Usual techniques such as reverse transcription polymerase chain reaction (RT-PCR) require labeling and enzymatic amplification to reach detectable signals. Instead, it is quite straightforward to arrange ssDNA probes to form simple (similar to barcode) patterns on flat DNA origamis. After binding of mRNAs to the probes, the origamis can be deposited on mica surfaces and imaged using atomic force microscopy (AFM). The detection of mRNA relies on the different mechanical properties of ssDNA and double-stranded DNA (dsDNA). Whereas ssDNA is quite elastic (almost invisible to AFM), dsDNA is much stiffer (persistence length ~50 nm) and can be visualized by AFM. As a consequence, the binding of mRNAs translates to a change in the AFM image, which can be resolved at the single origami scale. Interestingly, the probe–signal binding turns out to be quite inhomogeneous all over the origami, the binding being stronger close to the borders. The same technique allows simultaneous detection of several mRNAs, down to the 100 pM range. The authors of Ref. [21] concluded that the main limitation of this technique was related to the possibility of handling a small amount of DNA origami tiles onto a tiny spot for AFM imaging. For instance, 1 nL of a 1 pM solution of tiles could bring the sensitivity of the method down to ~1000 molecules, adequate to characterize the genetic content of single cells.

6.3.2 Arranging Enzymatic Cascades in Vitro

Several groups have reported successful assembly of enzymatic networks on DNA double helices [22, 23] or in 2D DNA or ARN periodic arrays [24, 25]. DNA origamis offer additional possibilities for patterning, as reported in Ref. [26]. In this work, the authors considered coupling between glucose oxidase (GOx) and horseradish peroxidase (HRP). Both enzymes were conjugated (through hystidines) with 20 nt long ssDNA so that the assembly with the DNA origami proceeded simply by Watson–Crick hybridization with extended staples. The efficiency of this binding mode depended on the distance between the two enzymes, being almost perfect for distances >60 nm and much lower (45%) for closer points of attachment (~10 nm). The enzymatic activity was diffusion controlled and remained constant down to an enzyme interspacing of 10 nm. At this gap value an enhancement of more than 15 times the unbound enzyme activity was observed. The transport mechanism in the close (~10 nm) position was no more dictated by free 3D diffusion, but surface-limited diffusion (the reaction intermediates transfer is mediated by the hydration layer of both enzymes). The authors were able to confirm this point by adding a third protein between GOx and HRP. This new protein had no enzymatic role but provided an additional hydration layer to enhance H_2O_2 transport between GOx and HRP.

Other modes of binding have been explored: reversible aptamer binding [27, 28], antibody–antigen interactions [29, 30], and the more stable biotin–streptavidin binding [31]. A more general method, leading to an *orthogonal protein decoration* of DNA origami, has been described in Ref. [32]. These authors used self-labeled fusion proteins with Snap-tags [33] and Halo-tags [34]. These tags allow the specific binding to alkylamino-modified oligonucleotides on the DNA origami. In Ref. [32], the authors pushed further the method to obtain Janus origamis, namely, planar DNA origamis decorated on both sides, which allowed directional immobilization of these planar structures.

6.3.3 Arranging Enzymatic Cascades in Vivo

Previous sections have shown a large variety of structure-based generalizations of Rothemund's original idea [1]. All these extensions share a common point: the need for an annealing process for the assembly of the platform. Even though recent findings [35] tend to show that, even for 3D origamis, the annealing process can be significantly shortened compared to the original experimental protocols, rather high temperatures (~60°C) are needed at some point. From this point of view, DNA origamis seem incompatible with natural conditions of cell metabolism. Other unfolding agents (such as urea [36]) could be used isothermally, with the same constraints regarding the cell metabolism.

On the contrary, nucleic acid arrays are easier to devise, as they only need two types of strands to form and can do so isothermally. In Ref. [25], the authors showed that RNA scaffolds can be assembled inside a living cell. For this, they genetically modified bacteria, expressing a few types of RNA strands, with aptamer domains that bind PP7 and MS2 fusion proteins [37]. Inspired by previous DNA design of arrays [37, 38], Delebecque et al. devised 1D and 2D RNA arrays that could bind two chimeras containing, respectively, [FeFe]-hydrogenase and ferredoxin. Coexpression of these two proteins in the cell, together with transcription of a self-assembling RNA 2D network, led to an efficient hydrogen-producing pathway [39]. The formation of arrays actually occurs in two steps: First, the strands fold in such a way that part of the sequence involved in the polymerization step (array formation) is hidden in a stem loop. When two of these folded strands bind to form a dimer, the polymerization sequence is activated, leading to the formation of chains of dimers. Activity measurements showed that hydrogen production could proceed almost 50 times faster in the presence of the ARN scaffold than in solution.

6.4 Challenges and Future Directions

In the past years, DNA-based bottom-up approaches have proven to be particularly attractive for the design of complex molecular nanostructures. DNA origamis have enlarged the perspectives in this field by providing a generic approach to generate increasingly complex constructions. In particular, we have discussed the possibility to decorate in a reliable and specific way these platforms with proteins and nucleic acids. First experiments have shown that bound enzymatic complexes remain active. Depending on the relative distance between enzymes, this activity can be enhanced by at least 1 order of magnitude as compared to free enzymes in solution. Remarkably, the same activity enhancement has been observed in in vivo experiments for couples of enzymes attached to an ARN array.

The control on the proximity between two active biomolecules has proven to be of importance in other contexts than enzymatic activity. Reference [40] considered the efficiency of thrombin aptamers when bound to a small DNA platform formed by two intertwined ssDNAs. This nanostructure allowed optimization of spacing and orientation of the aptamers to obtain an eightfold increase of the anticoagulant activity. The authors of Ref. [41] assembled in a tetrahedral DNA nanostructure [42] a model antigen (streptavidin) with an adjuvant (mouse-specific ODN-1862) to form a synthetic vaccine complex that was shown to induce strong and long-lasting antibody responses.

It seems clear that these first attempts will be followed by many other experiments involving a limited number of biomolecules and, more generally, as a tool in single-molecule biophysics. In this direction, the pioneering AFM observation of conformational switches in G-quadruplexes [43] or the regulation of DNA methylation [44] under tension is worth mentioning. Basic problems to be solved in the future [45] include minimizing the interaction with the AFM substrate (punched origamis, as those in Refs. [43, 44], are one possibility), increasing the number of methods to attach oligonucleotides to specific residues of the proteins, and improving the orientation control of the biomolecules. This third task could be achieved by multiple attachments to DNA origamis or (as partly tried in Ref. [31]) by 3D cavities that constrain the incoming molecules through steric interactions.

Vectorizing these DNA nanostructures remains a largely open challenge. It has been shown that DNA origamis are very stable against digestion by nucleases [46] or inside cell lysates [47]. Vectorizing these structures inside the cells in an efficient way will probably rely on the addressability of their surfaces, to which molecules such as antibodies or hormones could be easily grafted [48].

Acknowledgments

This work has been partially funded by Video Imaging of Biological and Bioinspired Nanosystems (VIBBnano) and Agence Nationale de la Recherche (ANR).

References

1. Rothemund, P. W. K. (2006) Folding DNA to create nanoscale shapes and patterns, *Nature*, **440**, 297–302.
2. Kallenbach, N., Ma, R., and Seeman, N. (1983) An immobile nucleic-acid junction constructed from oligonucleotides, *Nature*, **305**, 829–831.
3. Winfree, E., Liu, F. R., Wenzler, L. A., and Seeman, N. C. (1998) Design and self-assembly of two-dimensional DNA crystals, *Nature*, **394**, 539–544.
4. Schulman, R., and Winfree, E. (2007) Synthesis of crystals with a programmable kinetic barrier to nucleation, *Proc. Natl. Acad. Sci. U. S. A.*, **104**, 15236–15241.
5. Rothemund, P. W. K., Papadakis, N., and Winfree, E. (2004) Algorithmic self-assembly of DNA Sierpinski triangles, *PLOS Biol.*, **2**, 2041–2053.
6. Rothemund, P. W., Ekani-Nkodo, A., Papadakis, N., Kumar, A., Fygenson, D. K., and Winfree, E. (2004) Design and characterization of programmable DNA nanotubes, *J. Am. Chem. Soc*, **126**, 16344–16352.
7. Arbona, J. M., Aime, J. P., and Elezgaray, J. (2013) Cooperativity in the annealing of DNA origamis, *J. Chem. Phys.*, **138**, 015105.
8. Arbona, J. M., Elezgaray, J., and Aime, J. P. (2012) Modelling the folding of DNA origami, *EPL*, **100**, 28006.
9. Arbona, J. M., Aime, J. P., and Elezgaray, J. (2012) Modeling the mechanical properties of DNA nanostructures, *Phys. Rev. E Stat. Nonlin., Soft Matter Phys.*, **86**.
10. Sobczak, J. P. J., Martin, T. G., Gerling, T., and Dietz, H. (2012) Rapid folding of DNA into nanoscale shapes at constant temperature, *Science*, **338**, 1458–1461.
11. Andersen, E. S., Dong, M., Nielsen, M. M., Jahn, K., Subramani, R., Mamdouh, W., Golas, M. M., Sander, B., Stark, H., Oliveira, C. L., Pedersen, J. S., Birkedal, V., Besenbacher, F., Gothelf, K. V., and Kjems, J. (2009) Self-assembly of a nanoscale DNA box with a controllable lid, *Nature*, **459**, 73–76.

12. Han, D., Pal, S., Nangreave, J., Deng, Z., Liu, Y., and Yan, H. (2011) DNA origami with complex curvatures in three-dimensional space, *Science*, **332**, 342–346.

13. Douglas, S. M., Dietz, H., Liedl, T., Högberg, B., Graf, F., and Shih, W. M. (2009) Self-assembly of DNA into nanoscale three-dimensional shapes, *Nature*, **459**, 414–418.

14. Zadegan, R. M., Jepsen, M. D., Thomsen, K. E., Okholm, A. H., Schaffert, D. H., Andersen, E. S., Birkedal, V., and Kjems, J. (2012) Construction of a 4 Zeptoliters Switchable 3D DNA Box Origami, *ACS Nano*, **6**, 10050–10053.

15. Douglas, S. M., Bachelet, I., and Church, G. M. (2012) A logic-gated nanorobot for targeted transport of molecular payloads, *Science*, **335**, 831–834.

16. Douglas, S. M., Marblestone, A. H., Teerapittayanon, S., Vazquez, A., Church, G. M., and Shih, W. M. (2009) Rapid prototyping of 3D DNA-origami shapes with caDNAno, *Nucleic Acids Res.*, **37**, 5001–5006.

17. Green, L. S., Jellinek, D., Jenison, R., Östman, A., Heldin, C. H., and Janjić, N. (1996) Inhibitory DNA ligands to platelet-derived growth factor B-chain, *Biochemistry*, **35**, 14413–14424.

18. Tang, Z., Shangguan, D., Wang, K., Shi, H., Sefah, K., Mallikratchy, P., Chen, H. W., Li, Y., and Tan, W. (2007) Selection of aptamers for molecular recognition and characterization of cancer cells, *Anal. Chem.*, **79**, 4900–4907.

19. Huang, Y. F., Shangguan, D., Liu, H., Phillips, J. A., Zhang, X., Chen, Y., and Tan, W. (2009) Molecular assembly of an aptamerdrug conjugate for targeted drug delivery to tumor cells, *ChemBioChem*, **10**, 862–868.

20. Perham, R. N. (2000) Swinging arms and swinging domains in multifunctional enzymes: catalytic machines for multistep reactions, *Annu. Rev. Biochem.*, **69**, 961–1004.

21. Ke, Y., Lindsay, S., Chang, Y., Liu, Y., and Yan, H. (2008) Self-assembled water-soluble nucleic acid probe tiles for label-free RNA hybridization assays, *Science*, **319**, 180–183.

22. Niemeyer, C. M., Koehler, J., and Wuerdemann, C. (2002) DNA-directed immobilization of an artificial bienzyme consisting of NAD(P)H:FMN oxidoreductase and luciferase, *ChemBioChem*, **3**, 242.

23. Erkelenz, M., Kuo, C. H., and Niemeyer, C. M. (2011) DNA-mediated assembly of cytochrome P450 BM3 subdomains, *J. Am. Chem. Soc.*, **133**, 16111–16118.

24. Wilner, O. I., Weizmann, Y., Gill, R., Lioubashevski, O., Freeman, R., and Willner, I. (2009) Enzyme cascades activated on topologically programmed DNA scaffolds, *Nat. Nanotech.*, **4**, 249–255.

25. Delebecque, C. J., Lindner, A. B., Silver, P. A., and Aldaye, F. A. (2011) Organization of intracellular reactions with rationally designed RNA assemblies, *Science*, **333**, 470–474.
26. Fu, J., Liu, M., Liu, Y., Woodbury, N. W., and Yan, H. (2012) Interenzyme substrate diffusion for an enzyme cascade organized on spatially addressable DNA nanostructures, *J. Am. Chem. Soc.*, **134**, 5516–5519.
27. Chhabra, R., Sharma, J., Ke, Y., Liu, Y., Rinker, S., Lindsay, S., and Yan H. (2007) Spatially addressable multi-protein nanoarrays templated by aptamer tagged DNA nanoarchitectures, *J. Am. Chem. Soc.*, **129**, 10304–10305.
28. Rinker, S., Ke, Y., Liu, Y., Chhabra, R., and Yan, H. (2008) Self-assembled DNA nanostructures for distance dependent multivalent ligand-protein binding, *Nat. Nanotech.*, **3**, 418–422.
29. He, Y., Tian, Y., Ribbe, A. E., and Mao, C. (2006) Antibody nanoarrays with a pitch of approximately 20 nanometers, *J. Am. Chem. Soc.*, **128**, 12664–12665
30. Williams, B. A. R., Lund, K., Liu, Y., Yan, H., and Chaput, J. C. (2007) Self-assembled peptide nanoarrays: an approach to studying protein-protein interactions, *Angew. Chem.*, **119**, 3111–3114.
31. Numajiri, K., Yamazaki, T., Kimura, M., Kuzuya, A., and Komiyama, M. (2010) Discrete and active enzyme nanoarrays on DNA origami scaffolds purified by affinity tag separation, *J. Am. Chem. Soc.*, **132**, 9937–9939.
32. Saccà, B., Meyer, R., Erkelenz, M., Kiko, K., Arndt, A., Schroeder, H., Rabe, K. S., and Niemeyer C. M. (2010) Orthogonal protein decoration of DNA origami, *Angew. Chem., Int. Ed.*, **49**, 9378–9383.
33. Keppler, A., Gendreizig, S., Gronemeyer, T., Pick, H., Vogel, H., and Johnsson, K. (2003) A general method for the covalent labeling of fusion proteins with small molecules in vivo, *Nat. Biotechnol.*, **21**, 86-89.
34. Los, G. V., and Wood, K. (2007) The HaloTag: a novel technology for cell imaging and protein analysis, *Methods Mol. Biol.*, **356**, 195–208.
35. Sobczak, J. P., Martin, T. G. Gerling T., and Dietz, H. (2012) Rapid folding of DNA into nanoscale shapes at constant temperature, *Science*, **338**, 1458–1461.
36. Jungmann, R., Liedl, T., Sobey, T. L., Shih, W., and Simmel, F. C. (2008) Isothermal assembly of DNA origami structures using denaturing agents, *J. Am. Chem. Soc.*, **130**, 10062–10063.
37. Liu, H., He, Y., Ribbe, A. E., and Mao, C. D. (2005) Two-dimensional (2D) DNA crystals assembled from two DNA strands, *Biomacromolecules*, **6**, 2943–2945.

38. Liu, H., Chen, Y., He, Y., Ribbe, A. E., and Mao, C. D. (2006) Approaching the limit: can one DNA oligonucleotide assemble into large nanostructures? *Angew. Chem., Int. Ed. Engl.*, **45**, 1942–1945.

39. Agapakis, C. M., and Silver, P. A. (2010) Modular electron transfer circuits for synthetic biology: insulation of an engineered biohydrogen pathway, *Bioeng. Bugs*, **1**, 413–418.

40. Rangnekar, A., Zhang, A. M., Li, S. S., Bompiani, K. M., Hansen, M. N., Gothelf, K. V., Sullenger, B. A., and LaBean, T. H. (2012) Increased anticoagulant activity of thrombin-binding DNA aptamers by nanoscale organization on DNA nanostructures, *Nanomedicine*, **8**, 673–681.

41. Liu, X., Xu, Y., Yu, T., Craig, C., Liu, Y., Yan, H., and Chang, Y. (2012) A DNA nanostructure platform for directed assembly of synthetic vaccines, *Nano Lett.*, **12**, 4254–4259.

42. Zhang, C., Tian, C., Guo, F., Liu, Z., Jiang, W., and Mao, C (2012) DNA-directed three-dimensional protein organization, *Angew. Chem., Int. Ed. Engl.*, **51**, 3382–3385.

43. Sannohe, Y., Sannohe, Y., Endo, M., Katsuda, Y., Hidaka, K., and Sugiyama, H. (2010) Visualization of dynamic conformational switching of the G-quadruplex in a DNA nanostructure, *J. Am. Chem. Soc.*, **132**, 16311–16313.

44. Endo, M., Katsuda, Y., Hidaka, K., and Sugiyama, H. (2010) Regulation of DNA methylation using different tensions of double strands constructed in a defined DNA nanostructure, *J. Am. Chem. Soc.*, **132**, 1592–1597.

45. Pinheiro, A. V., Han, D., Shih, W. M., and Yan, H. (2011) Challenges and opportunities for structural DNA nanotechnology, *Nature Nanotechnol.*, **6**, 763–772.

46. Castro, C. E., Kilchherr, F., Kim, D. N., Shiao, E. L., Wauer, T., Wortmann, P., Bathe, M., and Dietz, H. (2011) A primer to scaffolded DNA origami, *Nat. Methods*, **8**, 221–229.

47. Mei, Q., Wei, X., Su, F., Liu, Y., Youngbull, C., Johnson, R., Lindsay, S., Yan, H., and Meldrum, D. (2011) Stability of DNA origami nanoarrays in cell lysate, *Nano Lett.*, **11**, 1477–1482.

48. Rothemund, Y., Buchwald, C., Georgiewa, P., Bohner, G., Bauknecht, H. C., Ballmaier, M., Klapp, B. F., and Klingebiel, R. (2011) Compulsivity predicts fronto striatal activation in severely anorectic individuals, *Neuroscience*, **197**, 242–250.

Chapter 7

Nanopatterning for Nanobiotechnologies: Emerging Methods Based on Soft Lithography and Directed Assembly

Aline Cerf,[a,b] Christophe Thibault,[a,b] Emmanuelle Trévisiol,[a,b] and Christophe Vieu[a,c]

[a]*CNRS, LAAS, 7 avenue du colonel Roche, F-31400 Toulouse, France*
[b]*Université de Toulouse, LAAS, 7 avenue du colonel Roche, F-31400 Toulouse, France*
[c]*Université de Toulouse, INSA, LAAS, 7 avenue du colonel Roche, F-31400 Toulouse, France*
acerf@laas.fr

Through miniaturization down to the molecular scale, the fabrication of innovative devices, systems, and chips dedicated to medical applications becomes possible. The expectations are very high in terms of diagnosis at the early stage of a disease, for new therapies and for increasing the capabilities of medical imagery. Another issue concerns the use of these technologies for the development of new experimental approaches for investigating living systems and for acquiring new knowledge in biology at the cellular and subcellular levels. Indeed, through micro- and nanotechnologies

Enzyme Nanocarriers
Edited by Daniela Cardinale and Thierry Michon
Copyright © 2016 Pan Stanford Publishing Pte. Ltd.
ISBN 978-981-4613-42-2 (Hardcover), 978-981-4613-43-9 (eBook)
www.panstanford.com

the manipulation and observations of DNA, proteins, and cells allow new experiments to be conducted in the perspective of investigating in a new way basic fundamental mechanisms of biology. At the core of these systems and methods, nanopatterning is a crucial step as it allows building the interface between the inert part of the device and the biomolecule or cell. During the last decade, many new processes dedicated to immobilize bioentities without denaturing them and with submicrometric spatial resolution have been developed. Among them soft lithography combines easy implementation, low cost, and high throughput, and when combined with capillary assembly, it turned out to be capable of single-molecule resolution or single-cell manipulation. In this chapter, we illustrate these achievements through some examples related to biopatterning, biodetection, and single-molecule manipulations and open perspectives in the field of patterning virus nanoparticles.

7.1 Introduction on a Vision of the Future of Nanobiotechnologies

During the last century, two major technologies have emerged and have deeply modified our common perception of science and technology and are shaping our society. The first one is related to the exploding field of semiconductor technology, starting in the 1950s with the last developments of VLSI circuits and the currently penetrating systems assembled under the umbrella of nanoelectronics. The second one is related to the development of molecular biology, starting with the discovery of the structure of DNA molecules in 1953 and the last developments of biotechnologies. Since 2000, many groups around the world are working in the perspective of combining these two blooming technologies. This new field, sometimes called *nanobiotechnologies*, has been the matter of funding in Europe (starting with the Network of Excellence of FP6 NanoToLife), the United States, and Asia. For most of the projects, applications and patents focusing on the combination of silicon technologies and biotechnologies, the dominant approach consists of using advanced nanoelectronic platforms fabricated massively using VLSI technology for biological or medical purposes. The industrial products, fruits of this kind of investigations, are already appearing in the market. It is clear today that multisensorial platforms, such as mobile phones, will be used for environment and health monitoring of people. However, this first generation of nanobiotechnology is only a convergence driven by applications of economic interests rather than a true hybridization of the two technological domains. One may then answer what the next technological step will be at the crossroads of nanoelectronics and biotechnologies. Our feeling is that the future generation of systems will incorporate, at the level of individual components, biological species (either molecular or cellular) and artificial devices (electrical, mechanical, or optical) fabricated by modern nanofabrication facilities. This perspective requires many advanced research dedicated to discover the new processes, materials, and technological bricks required for truly achieving hybrid bio-/artificial systems. We believe that these new processes will constitute the basis for future manufactured products arriving in the market at the horizon of 2025. The integration of bio entities is therefore a crucial starting point involving innovative techniques and processes, mixing biological entities with artificial micro- and nanoscale devices.

"Biointegrated" technologies generate a fascinating new technological domain where molecular assembly, nano-objects synthesized with "bottom up" methods, functional proteins, and bionanomachines will drastically change the already rich world of artificial systems. Two mainstreams for the integration of biology in nanotechnologies can be distinguished. The first one referred to as "nanotechnologies for biology" includes the generation of active devices capable of manipulating cells and biomolecules for biomarker detection or tissue engineering as well as modeling of molecular interactions and biological mechanisms. The second one referred to as "nanotechnologies from biology" relies on the fabrication of original nanodevices where the main functionality will be given by a bioentity such as DNA origami, DNA-based nanochannels, and bioassembled materials, to name a few.

Integrating bioentities on devices requires a method for locating cells or biomolecules on surfaces with high spatial accuracy, while preserving their biological functionality. This technological process is called biopatterning. The spatial resolution and the quality of the bioentities arrangement on the surface are of great importance for various applications. In this chapter, we will show why soft lithography and, in particular, microcontact printing (μCP) exhibit specific advantages for advanced biopatterning. This novel lithography method introduced by Whiteside's group [1] has become a well-established technique, capable of patterning in a highly parallel way a large variety of molecules on arbitrary surfaces for both research purposes and technical applications. Initially, the technique was proposed as a route for creating patterns of self-assembled monolayers (SAMs) [1–3] and was later extended to the patterning of biomolecules [4, 5] as well as functional molecules on surfaces using supramolecular interactions [6]. The transfer of these molecules along well-controlled features is achieved through the use of a patterned elastomeric stamp, typically fabricated from poly(dimethylsiloxane) (PDMS) [7] cast onto a lithographically patterned silicon hard master. This material turned out to possess unique advantages with respect to the printing process, such as its ability to exhibit conformal contact on rough surfaces and its chemical inertness.

In the first part, we will show how μCP can be used for the fabrication of DNA microarrays [8]. It has been shown that inking times and contact times of less than 30 seconds give high-quality and high-resolution microarrays. Moreover, it was found that the fluorescence signal emitted by DNA hybridization spots created by μCP was systematically higher compared to similar spots created by conventional tip deposition. To interpret the rapid and efficient adsorption of DNA molecules on a freshly cured PDMS stamp, the presence of PDMS fragments in the elastomeric stamp turned out to improve the quality of the transfer of the molecules [9].

In the second part, we will show that by biopatterning probe molecules at the nanoscale using soft lithography, innovative protein biochips can be produced. The combination of multiplexed nanoscale μCP and label-free optical detection using the principle of light diffraction has indeed been implemented for generating engineered glass slides for label-free analysis [10, 11].

In the last part, we will show how soft lithography can be used for investigating single bioentities. Arrays of single living bacteria [12, 13] and perfectly controlled arrays of single stretched DNA molecules for genetic analysis or medical diagnosis [14] can be produced by combining soft lithography with directed capillary assembly. Using similar principles, periodic bidimensional matrixes of gold nanoparticles [15] can be fabricated for different applications in the field of plasmonics.

7.2 Microcontact Printing of Oligonucleotides for Biochip Fabrication

In a pioneering paper, Lange et al. [16] showed that the μCP technique could be used to deposit DNA molecules with a chemically modified PDMS stamp surface enabling the DNA molecules to stick on the stamp. This functionalization step strongly restricted the speed of this technology, as it takes several hours from the conversion of the alkyl-terminated surface of the PDMS into an aminated surface to complete inking of the stamps prior to printing the target surface. Later, Thibault et al. [8] demonstrated that μCP could be directly used to fabricate DNA biochips without any surface modification of the stamp. These authors showed that inking and contact times of less than 30 seconds produce high-quality and high-resolution arrays by μCP. According to this new variant of the process, the stamp is simply inked with the molecules of interest (DNA probes), dried under a nitrogen stream, and then printed manually onto the substrate surface (Fig. 7.1). The reason for the rapid and efficient adsorption of DNA molecules onto untreated hydrophobic PDMS stamps was not clear, and we chose to investigate the possible role of PDMS fragments present in the elastomeric stamps after crosslinking in this behavior. We thus investigated an extraction process by Soxhlet cleaning, which eliminates all traces of contamination. This cleaning process led us to evaluate its impact on the quality of microcontact printed patterns of DNA molecules.

Indeed, when a PDMS stamp contacts the surface of the substrate it rapidly releases some molecules exhibiting a hydrophobic character. By looking at the wettability of the substrate's surface it is possible to obtain a macroscopic signature of this contamination process. Starting with a hydrophilic substrate of glass or oxidized silicon, we observed a significant change of the contact angle at the substrate–water interface after contact times as short as a few seconds with flat PDMS stamps. To investigate the level of contamination and the evolution of the residues on the substrate's surface after printing we systematically measured the evolution of the water–substrate contact angle over time after printing with flat PDMS stamps for different contact durations, ranging from 15 seconds to 5 minutes. A typical curve is presented in Fig. 7.2 (blue curve, diamonds). We can observe in Fig. 7.2 that the evolution of the hydrophobicity of the surface after PDMS contamination exhibits two distinct stages. A very rapid increase during the very first minutes from a value α_i corresponding to the contact angle of the initial hydrophilic oxidized surface to an intermediate value close to 70° is followed by a much slower evolution that reaches a steady state after a few hours, corresponding to a contact angle α_f greater than 90°. This characteristic profile was observed for different samples and different PDMS preparation conditions.

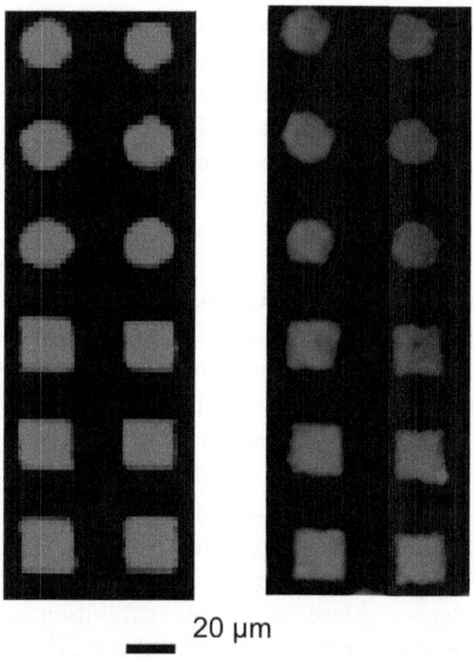

Figure 7.1 μCP of oligonucleotides and PCR fragments. Fluorescent images of typical micrometric printed features. The PDMS stamp was incubated for 30 sec with a 20-mer oligonucleotide of the *HSP12* yeast gene (left side) or with a 500 bp PCR fragment (dsDNA) of the yeast *HSP12* gene (right side) and then put in contact manually for 15 sec onto a dendrislide. Hybridization was carried out with the complementary *HSP12* Cy5-labeled oligonucleotide.

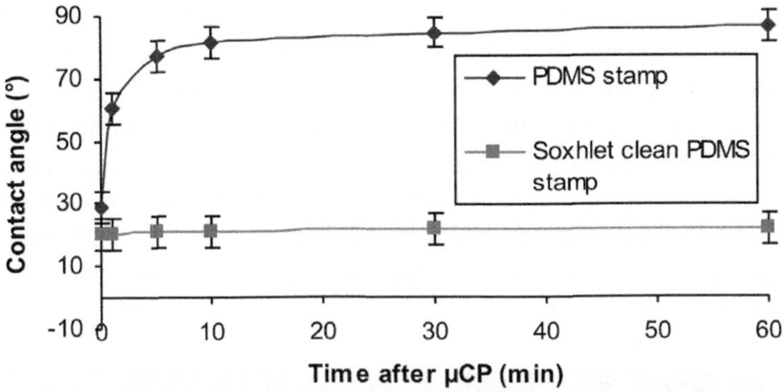

Figure 7.2 Evolution of the water–substrate contact angle (glass substrate) over time after printing with flat PDMS stamps. The curve named "PDMS stamp" represents the evolution over time of the water–substrate contact angle after printing with flat standard PDMS Sylgard 184® stamps. The second curve named "Soxhlet cleaned PDMS stamp" represents the evolution over time of the water–substrate contact angle after printing with flat standard PDMS Sylgard 184® stamps previously cleaned with the Soxhlet device.

After cleaning PDMS stamps, it was not possible to observe any trace of contamination after printing. Indeed, the water–substrate contact angle does not significantly increase after contacting a cleaned stamp with silicon or glass surfaces (Fig. 7.2, red curve, squares). The evolution over time of the water–substrate contact angle cannot be distinguished from the slightly progressive increase related to the unavoidable contamination of the surface at ambient atmosphere.

Having found a reliable method for eliminating the presence of siloxane fragments inside and at the surface of PDMS stamps, we studied the impact of this silicone oil on the µCP process itself. Mainly we wanted to know if the presence of these contaminating molecules is beneficial or detrimental with respect to the quality and characteristics of a printing process. We have investigated the impact of the cleaning of PDMS stamps on the µCP process of biomolecules and specifically DNA molecules. A typical image of fluorescent single-stranded DNA spots created by µCP is shown in Fig. 7.3. In this image we can observe the red fluorescence emitted by the DNA molecules used for µCP (20-mer labeled with Cy5 commercial fluorophores). On the left side, the printing process was achieved with an uncleaned

PDMS stamp containing micrometric features compatible with fluorescence scanner characterization. On the right side, the printing process was achieved in the exact same conditions, the same type of stamp, except this time the stamp underwent a cleaning process before μCP. The same experiments were repeated a large number of times and systematically indicated that the fluorescence emitted by the spots printed with an uncleaned stamp is significantly higher than the signal measured on spots printed with a PDMS stamp free of silicone fragments.

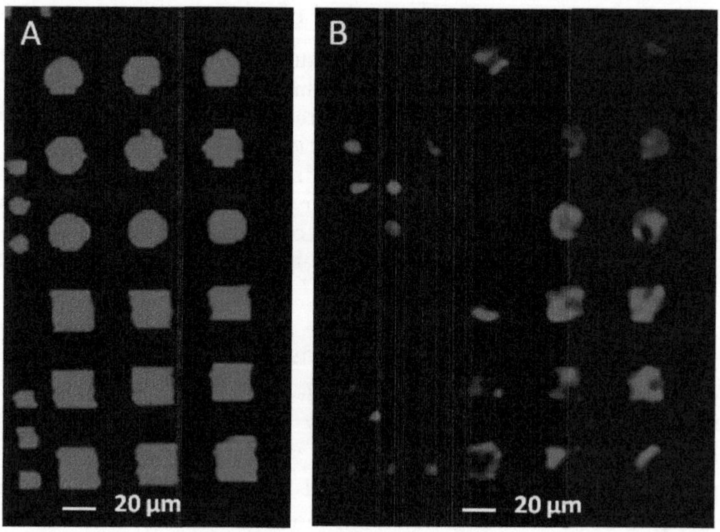

Figure 7.3 Fluorescence image of DNA spots printed by μCP on a glass slide. The image on the left (A) corresponds to a printing process achieved with an uncleaned PDMS stamp containing micrometric features compatible with fluorescence scanner characterization. The image on the right side (B) corresponds to a printing process achieved in the exact same conditions but with a Soxhlet-cleaned PDMS stamp.

The difference in the emitted signal was quantitatively recorded for different concentrations of DNA molecules in the inking solution. In addition to the fluorescence intensity, we also studied the homogeneity of the fluorescence signal across individual spots. Whatever the concentration in the inking solution is, the fluorescence from the spots printed with an uncleaned stamp is approximately four times higher than the fluorescence from spots printed with a cleaned PDMS stamp. Moreover, the fluctuations of the fluorescence signal are negligible for uncleaned stamps (1%–2%), while they can be as large as 15%–51% in the case of spots printed with a cleaned stamp. In the case of cleaned stamps, fluorescence become more heterogeneous as the concentration of the inking solution is reduced. This surprising result shows that in the case of DNA printing, the presence of PDMS fragments in the elastomeric stamp improves the quality of the molecules' transfer. We also observed this behavior after hybridization. In another experiment, the printed single-stranded DNA probes are not fluorescent and hybridization with a complementary fluorescent oligomer is performed. This experiment mimics a real DNA microarray assay. In Fig. 7.4, we have plotted the intensity of the fluorescence signal measured on 50 μm hybridization spots.

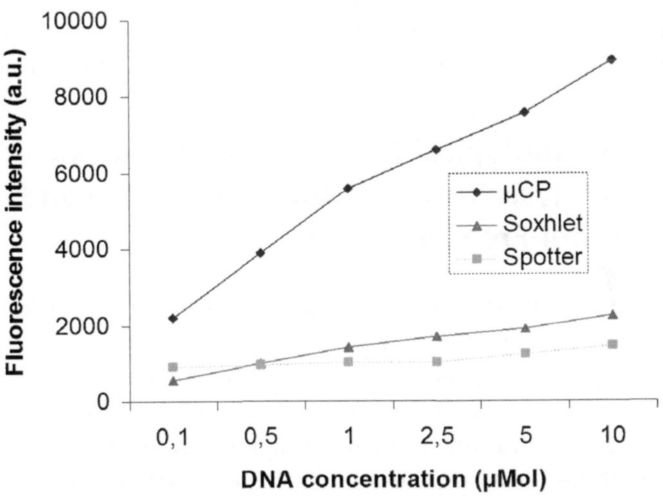

Figure 7.4 Comparison of the fluorescence intensity measured after DNA hybridization on DNA spots patterned with a conventional metal pin spotter (green curve, squares), by μCP using a PDMS stamp not cleaned after curing (blue curve, diamonds), and by μCP using a PDMS stamp cleaned during 12 hours with the Soxhlet system (red curve, triangles), for different concentrations of probe molecules. Each experimental data point corresponds to an average value of over hundreds of spots. The DNA probe molecules (20-mers) were spotted over a glass slide chemically treated with a dendrimer layer (see Ref. [1] for details).

We compare the case of probe molecules printed with a metal pin arrayer conventionally used for DNA microarray fabrication with DNA spots printed by μCP either with an uncleaned stamp susceptible to contaminating the surface with PDMS fragments or with a cleaned PDMS stamp, giving rise to DNA patterns free of silicone residues. In good agreement with the experiment reported in Fig. 7.3, we observe that whatever the concentration of DNA probe molecules is, the fluorescence intensity measured on spots printed with an uncleaned stamp are significantly higher than the signal measured on spots printed with a clean stamp or deposited by a metal pin. This experiment reproduces the results published by Thibault et al. [8] but gives an additional indication on the mechanism underlying the increasing fluorescence emissivity of DNA hybridization spots printed by μCP. Indeed, when using a cleaned PDMS stamp for printing the probe molecules, the fluorescence signal is not increased. This clearly demonstrates the role of PDMS fragments on the quality of the adsorption of DNA molecules on the stamp and/or their role for the packing of a high density of accessible probes on the surface.

7.3 Technological Platform for Molecular Analysis Based on Nanopatterning and Label-Free Optical Detection

In the field of nanomedicine and more specifically the field of oncology, nanotechnologies have been announced to deeply impact drug delivery processes and analytical devices for detecting various kinds of biomarkers with ultrahigh sensitivity [17]. After several demonstrations in research laboratory environments confirming these promises, these technologies based on various nano-objects, such as nanoparticles and nanowires, now face the problem of performance variabilities. Reliability will be a central issue in this field before some of these innovative devices become standard tools in cancer research. The reliability of medical diagnosis based on a molecular detection at the femtomolar level remains to be established in real conditions and real samples coming from patients. In this field, ultrahigh sensitivity conflicts dramatically with confidence and discriminating capabilities to avoid the multiplication of false positives.

Another aspect of the potential benefits of nanotechnology for medical diagnosis based on molecular analysis is the possibility to reduce drastically the cost of biochip technology. This improvement would render possible the availability of such molecular analysis at the level of genes or proteins for a large number of patients. Here the bottleneck, which is surpassed through nanotechnologies, is much more economical than cognitive. In this direction we have investigated a method for producing biochips (DNA or proteins) at low cost using soft lithography. Moreover, by exploiting the ability of this method to pattern probe molecules at the nanometric level, we have established a specific design of patterns enabling us to perform biodetection without any fluorescent labeling.

In any biochip technology, the ambition is to detect on a solid support a large number of possible specific interactions between probe molecules deposited on the surface, at registered positions, and target molecules present in the solution being analyzed at an unknown concentration. As for current DNA microarrays, through the precise knowledge of the sequences deposited on the matrix of spots, the detection of specific genes expressed in the analyte is rendered possible through the optical reading of the chip, as soon as target molecules are labeled with light-emitting fluorophores. This technology is very well established for DNA molecules where hybridization events are detected and has also been developed for mapping proteins by using principally specific antibodies as probe molecules. However, the cost of the method prevents its implementation for routine medical analysis. Today, the detection of multiple cancer biomarkers using protein biochips would probably have a major impact for medical diagnostics, but they are not employed due to their prohibitive cost of implementation. Only research activities can benefit from this technology. This high cost is mainly caused by three aspects: (i) the difficulty to deposit multiple probe molecules on a surface (typically a glass slide), which requires either expensive automatic robots or advanced photolithography methods; (ii) the requirement of labeling target molecules with fluorophores for enabling fluorescence detection to be performed; and (iii) the necessity to use large quantities of probe molecules, which can be expensive, due to the micrometric dimensions of the spots on the chip and the detection limit of fluorescence signals, which requires a given quantity of interactions in order for the fluorescent signal to be discriminated easily from the background noise. A new biochip platform at low cost should therefore combine a low-cost deposition method of multiple biological spots on a glass slide, a label-free detection method, and an improved limit of detection. The solution we have proposed relies on the use of soft lithography for the biopatterning process, a new label-free detection method based on the detection of diffracted optical beams, and the generation of nanoscale patterns, which enable a drastic reduction of the detection limit compared to fluorescence.

7.3.1 Deposition of Multiple Probe Biomolecules Using Microcontact Printing and Macrostamps®

During μCP, the elastomeric stamp is inked with a solution containing the molecules to be deposited and is gently brought into contact with the substrate. Ideally, the ink molecules are delivered to the surface only in the contact areas corresponding to the protrusions of the stamp. The method has proven to be able to pattern molecular features with a spatial resolution at the nanoscale [18, 19], can be implemented with limited equipment (manual printing works perfectly), and relies on very low cost material (PDMS). Of course, the silicon master used for creating the PDMS elastomeric replica requires advanced lithography methods for being fabricated [20], but a silicon master can be used a large number of times ($>10^3$), and each stamp can be regenerated for several successive prints. In Section 7.2 of this chapter, we have shown that, contrary to previous publications, the print could be achieved without any surface treatment of the PDMS stamp. The fluorescence signal arising from these printed spots turned out to be systematically much stronger compared to conventional spots deposited by metal pins due to a better organization of the DNA probes at the chip surface when deposited by printing. This clear advantage enabled us to use significantly smaller concentrations of probe molecules. Similar improvements were also observed for protein chips. In summary, we came rapidly to the conclusion that μCP could be used for high-quality deposition of biomolecules on glass slides for biochip applications. However, one major bottleneck remained to be solved for real useful applications. Indeed, one severe limitation of this technique is that only one type of biomolecule can be printed at a time because generally the stamp is inked everywhere with the same biomolecule. To be able to multiplex this patterning method, two methods have been developed. The first one called one-step-multiple μCP (OSM-μCP) [21, 22] uses a specific multilevel PDMS stamp allowing the patterning of two different molecules in one step. This original method, based on the elastomeric properties of a PDMS stamp, can print self-aligned patterns of two different molecules by exerting homogeneous pressure at the back of the stamp. During the controlled collapse of the stamp, the different levels inked with different molecules contact the surface sequentially. This method, however, cannot be used to increase the number of different molecules delivered over a large surface as the topographical structure of the stamp required to print *N* different molecules would require *N* different levels of PDMS protrusions and therefore *N* different depths etched in the silicon original master. A second methodology for printing several molecules in one step has thus been developed and patented. This multiplexing method targets the deposition of 100–1000 different molecules in one step [23]. For this specific purpose we have fabricated specific hierarchical stamps, called Macrostamps®, containing millimetric posts arranged in a periodic array compatible with titration plates (Fig. 7.5). Through a very simple molding procedure, each of these posts includes micro- and nanoscale features on its surface. Then, by inking in a single step each post with a different ink through a titration plate, it becomes possible to print different inks in one step and simultaneously obtain nanoscale patterns of many different biomolecules at a time. Our latest results show that without any trace of contamination up to 800 different biological inks can be "nanoprinted" in one step.

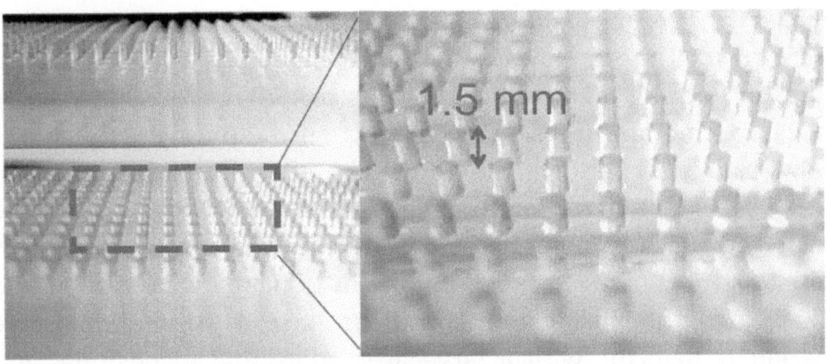

Figure 7.5 Photographies of a large multilevel PDMS stamp compliant with the format of a 1536 titration plate. Each of the protruding millimetric posts contains micro-/nanoscale features that cannot be seen at this magnification.

The validation of these multilevel stamps has been achieved using fluorescent inks and DNA probe molecules. The results we have obtained show that reliable printing can be obtained over a few hundreds of posts in one contact printing step and that no trace of contamination can be found from one post to the other.

7.3.2 Diffraction-Based Biosensing Using Nanoscale Periodic Gratings of Probe Molecules

Benefitting from the ability of μCP to pattern biomolecules along nanoscale features of any shape and dimension, we have investigated a new optical principle of detection that does not rely on any labeling method. The concept consists of arranging the probe molecules into periodic gratings able to efficiently diffract incident monochromatic light into various orders of diffraction. Upon specific interaction of these probe biomolecules with target molecules incubated in an analyte, the intensity of the diffracted light is modified, and this variation can be used for quantitative label-free biodetection (Fig. 7.6). This principle is implemented to finally build a scanner, which enables measurement of the diffracted intensity by each array, over a full biochip where several thousands of gratings are generated by multiplexed μCP. This diffractive scanner aims at replacing the conventional fluorescence scanner in current biochip technology.

Figure 7.7 shows an example of a periodic grating of proteins obtained by multiplexed μCP that is used for diffraction-based sensing. A full simulation of the diffractive behavior of these molecular gratings upon biomolecular interactions has been performed. The simulations show that biodetection by diffraction is a performing method that permits to differentiate, in a single step, specific and nonspecific interactions. Extensive simulations have then been performed with the aim of analyzing the influence of all the parameters. Some suggestions to choose the controllable experimental parameters would be normal incidence, wavelength as short as possible, small period but slightly superior to the wavelength, dimension of the gratings superior to four times the period p, detector with a large area adjustable in position, and linewidth near $l/p = 0.5$.

We have validated our detection method for protein–antibody interactions, and preliminary results have been obtained for DNA/

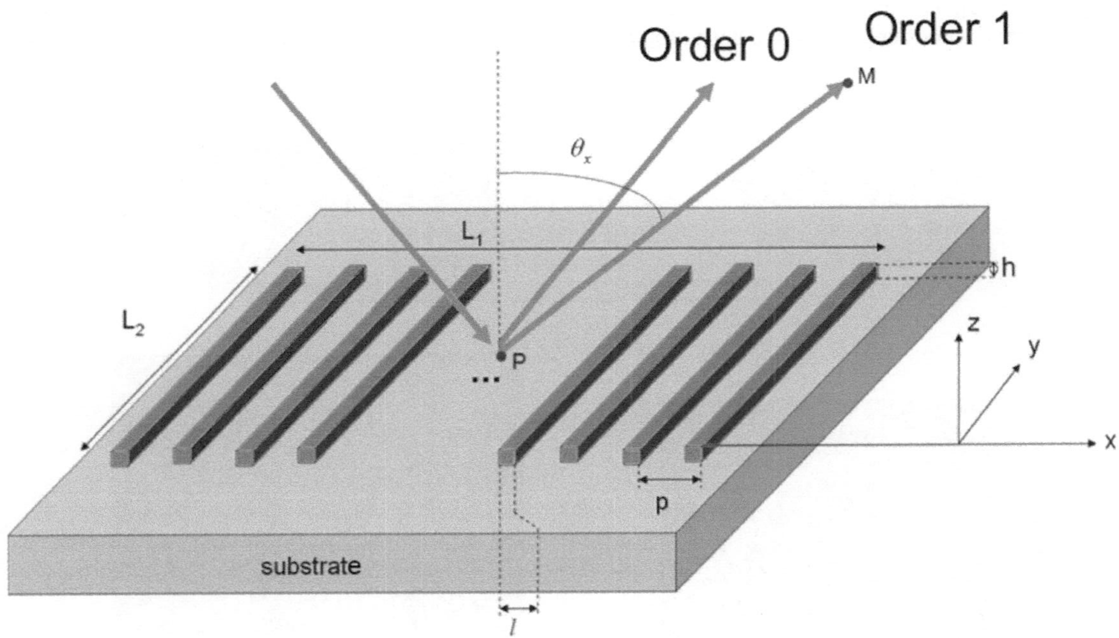

Figure 7.6 Principle of diffractive label-free biosensing. Probe molecules are printed by μCP into a periodic grating of lines. The period p is on the order of 1 μm, and the linewidth is submicrometric. An incident laser beam (red arrow) is efficiently diffracted into various orders of diffraction. The first-order beam is used for detecting specific probe–target interactions through a simple measurement of the diffracted intensity.

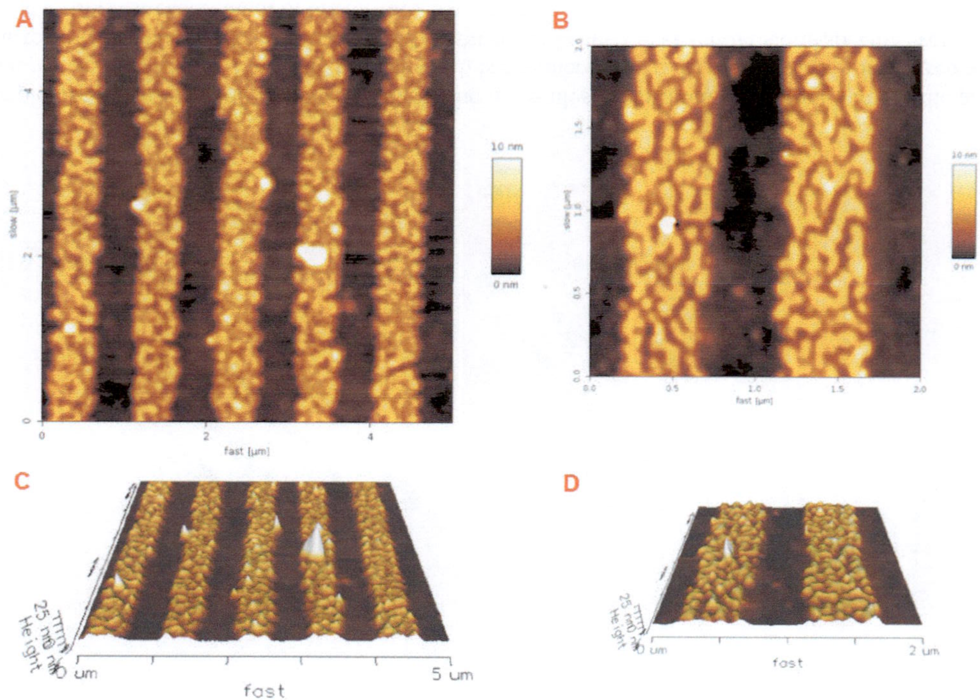

Figure 7.7 Atomic force microscopy (AFM) images of a nanoscale periodic grating of proteins (antibodies) fabricated by μCP used for optical label free detection of an immunoassay. The period of the grating is 1μm and the linewidth is 500 nm.

DNA hybridization detection [10, 11]. The same methodology has also been used successfully for the optical label-free detection of molecular spin crossover complexes [24].

7.4 Single Object Investigations Using Soft Lithography

Soft lithography and directed assembly are attractive and straightforward techniques capable of patterning and nano-ordering over large areas at low cost. To study single objects, the methodology we developed is declined following two strategies: either we combine soft lithography with a simple incubation technique, or we combine soft lithography with directed capillary assembly. Both approaches allow us to tightly control the assembly processes to direct the positioning and precise ordering of the objects of interest.

7.4.1 First Strategy: Surface Patterning

In the case of cells, we developed a fast, simple, and reproducible procedure for isolating and depositing individual living cells onto a substrate along well-defined and precisely registered micrometric patterns [12, 13, 25]. During the process, the species are not extracted from their natural media so there is no risk of damaging them. The approach consists of chemically engineering the surface with hydrophobic/hydrophilic contrast to induce and localize single-cell immobilization events. The principle of the methodology we perform for assembling cells is depicted in Fig. 7.8. It is based on the combination of conventional μCP with a simple incubation technique. In practice, the PDMS stamp with topographic structures is used to print the first ink, in this case a hydrophobic molecule on the surface. Then, by incubating the second ink, this time a hydrophilic molecule, the hydrophilic molecules fill out the empty patterns. After incubating a droplet of cells in suspension on this engineered surface, we obtain a selective adsorption of cells on the local chemical patterns producing highly ordered arrays of single living cells with a success rate close to 100% and over a 1 cm^2 area. By tuning the chemistry of the surface we are able to modulate the binding strength of cells onto the patterns. This allows the technique to be flexible as concerning the choice of the molecules used to graft cells either electrostatically or covalently to the surface by using untreated or chemically treated cells, respectively. The molecule used mainly influences the strength of the cells' binding onto the surface rather than their localization. In fact, only the surface microstructuration is responsible for the cells' trapping, so this trapping occurs whatever the affinity of cells with the surface is. By extension, this methodology could be a basic building block for interfacing living cells with artificial microsystems or for generating arrays of any kind of microorganisms by simply using the appropriate pattern functionalization.

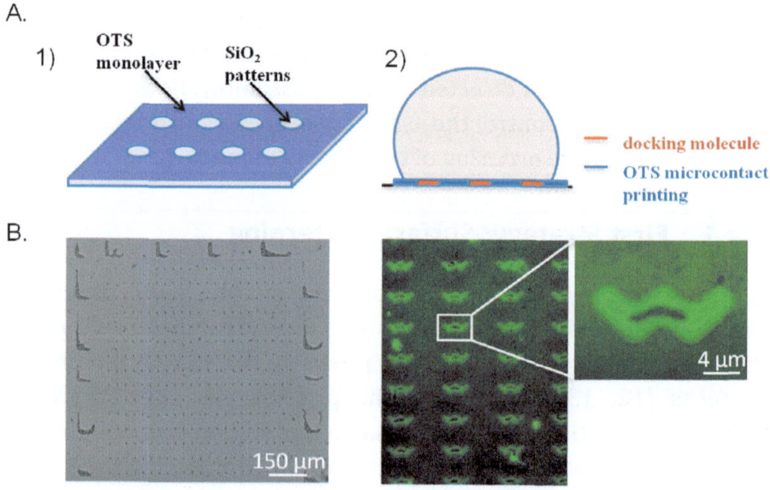

Figure 7.8 Chemically engineered substrates used to trap individual bacterial cells along the chemical patterns so defined (biopatterning). (A) Octadecyltrichlorosilane (OTS) µCP microcontact printing (1) followed by a functionalization of the bare patterns by incubation with docking molecules (2). (B) On the left, a bright field image of highly ordered arrays of single living bacteria, and on the right, fluorescence image of the same array. Here the patterns are functionalized with streptavidin molecules and bacteria are biotinylated so that they covalently bind to the surface.

Many other possibilities exist for tuning cell behavior through surface patterning by µCP. Recently we were able to control the adhesion of living macrophages on a glass slide by simply printing extracellular matrix proteins such as collagen, fibrinogen, and fibronectine. Depending on the printed protein, macrophages were found to respond to the surface patterns by self-organizing their actin-rich structures called podosomes on, around, or outside the printed areas [26]. This patterning capability allowed us to investigate by AFM the dynamics of the assembly and disassembly of the podosomes [26] or the frustrated phagocytosis caused by the patterns [27]. In another investigation we have developed an optimized μCP process for patterning carbon nanotube–thin carpets [28] and demonstrated the ability of these patterns to localize neuronal human cells, guide their growth, and align their neurites into very well-defined directions [29].

7.4.2 Second Strategy: Use of Structure Topography

If we travel now from the micrometric scale down to the nanometric scale, we find other interesting objects to study, like DNA molecules. Our purpose in this study was to single out individual DNA molecules and stretch them into controlled arrays for applications in the fields of genetics. For this particular purpose we did not use this time chemically engineered substrates but structured PDMS stamps as templates with cavities that would allow trapping of the objects of interest inside each cavity. In fact, in this case, we combine two techniques, soft lithography and directed capillary assembly. The directed assembly setup consists of a motorized translation stage where we place a PDMS stamp with micro- or nanostructures. Above this stamp we set a fixed glass slide at a distance of 1 mm, and we put in between a certain volume of the colloidal solution of the object to deposit. The liquid meniscus is then moved over the structured PDMS stamp at a controlled velocity (Scheme 7.1). As the meniscus encounters the stamp's topographic structures, it gets pinned for a certain time. Meanwhile, the objects undergo a convective flux and are directed toward the liquid front line where they are trapped by the capillary forces exerted inside each topographic cavity. When the elastic energy stored exceeds the pinning energy, the meniscus disrupts and releases the objects. By adjusting the displacement speed of the stage, the concentration of the colloidal solution, and the temperature of the PDMS substrate, and thus the evaporation rate of the solvent, we are able to control the number of objects to be trapped inside each micro- or nanowell of the structured PDMS

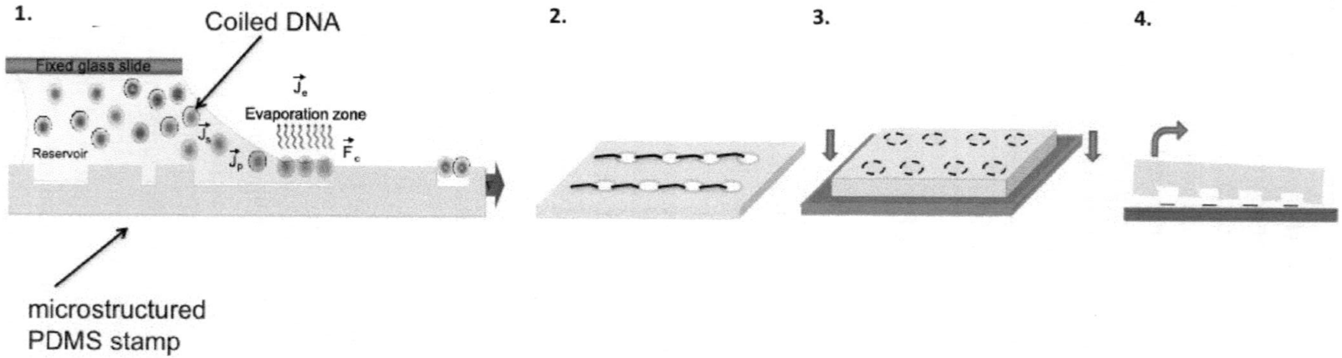

Scheme 7.1 Generation of single stretched DNA molecules arrays by directed assembly. (1) Capillary assembly of DNA molecules on a microstructured PDMS stamp. (2) Structured PDMS stamp with assembled stretched DNA molecules. (3) The assembly is then transfer-printed on an APTES-coated substrate. (4) The PDMS stamp is finally peeled away.

stamp with high placement accuracy. Once the objects are properly assembled under optimized experimental conditions, the objects are transferred onto a target substrate by simple contact printing.

In the case of DNA molecules, we use this particular methodology so as to trap individual coiled DNA molecules inside each well of a structured PDMS stamp and then stretch them by the disrupting of the meniscus. Here, the assembly is performed on a microstructured PDMS stamp at a relatively high velocity of 1 mm/sec, inducing almost simultaneously the selective trapping and stretching of single molecules at each cavity of the stamp. The resulting assembly of single stretched molecules is then transferred by μCP onto a freshly cleaved mica sheet or onto a glass slide coated by vapor deposition of 3-aminopropyltriethoxysilane (APTES) molecules. Figure 7.9 depicts one example of a highly regular array of stretched DNA molecules. This kind of result, under optimized conditions, could be obtained with a success rate close to 100% [14].

Figure 7.9 Highly organized chip of stretched DNA molecules.

The generation of these chips could be interesting for genome mapping or medical diagnostics through a fast and more efficient optical readout.

Another example of direct application is the assembly of nanoparticles [15, 30]. We explored the possibility to generate bidimensional matrixes of periodic sub-100 nm gold nanoparticles patterns by directed assembly. Small nanoparticles are the target of intense research due to their astonishing properties and their large window of applications in nanoelectronics, spintronics, optics, chemical sensing, etc. Assembling nanoparticles into well-ordered structures provides new nanostructures with remarkable and very distinct collective properties. In fact, by controlling their type, number, and precise arrangement, one can benefit from the properties they offer at that scale in the field of plasmonics in particular (surface-enhanced Raman scattering [SERS], metal-enhanced fluorescence [MEF], among others) and for the monitoring of specific interactions at low concentrations. By using the same technique as previously described, but this time by using a nanostructured PDMS stamp and by placing between the fixed glass slide and the mobile stamp a colloidal solution of 100 nm gold nanoparticles, we were able to generate matrixes of gold nanoparticles with different numbers of particles and different arrangements. In this case, the speed of the translation stage needs to be very low (1 μm/sec). Figure 7.10 shows the results obtained after directed assembly of gold nanoparticles on a nanostructured PDMS stamp and transfer-printing onto a silicon substrate. This time, accounting from the size of the nanoparticles, their irregular shape, and the fact that the resulting assembly is dry, the formed nanoparticles arrays were extremely difficult to transfer onto a solid substrate. We thus optimized a new method to transfer print an assembly from a PDMS stamp onto a substrate by performing the contact through a thin liquid film. In fact we deposit a droplet of solvent (deionized water or ethanol) on the target substrate coated by vapor deposition of APTES molecules and put the PDMS stamp with the assembled nanoparticles in contact. We put the overall in an oven at 100°C for a few minutes to evaporate all the solvent and peel the stamp off. As a result we obtain nice and regular nanoparticle patterns and the totality of the particles is transferred. Furthermore, we can observe from the dark-field micrographs that the assembled particles retain their optical activities (Fig. 7.10A), so more generally speaking, we could say that the assembly process does not alter the intrinsic properties of the objects [15].

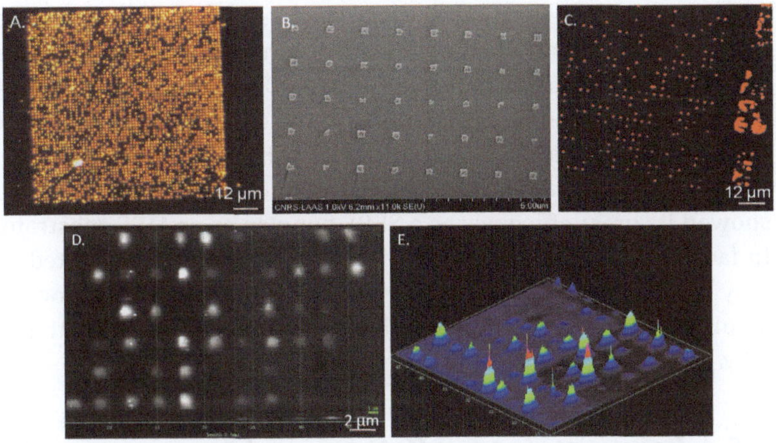

Figure 7.10 Dark field (A) and SEM (B) images of a periodic bidimensional matrix of 100 nm gold nanoparticles. (C) Confocal image of a controlled array of gold nanoparticles revealing an MEF effect. (D) Raman mapping of a controlled array of gold nanoparticles and (E) its equivalent in intensity levels revealing a SERS effect.

These encouraging results allowed us to imagine the further coupling of all these techniques to detect single objects or molecules at very low concentrations. In fact, we aimed at investigating the possibility to detect single objects via optical techniques. For enhancing Raman spectroscopic signals, we position the objects of interest in the close vicinity of noble metal nanocrystals (silver, gold). Such nanocrystals offer remarkable properties due to the localized surface plasmon resonances (SPRs) that induce, under the effect of an optical excitation with a frequency close to the SPR, a very intense local electric field enhancement nearby the surface. The interest of the SERS effect is that the enhancement factor can reach 10^{14}, sufficient enough to detect a single molecule [31]. The concept of ordered nanometer-sized metallic structures that allow controlling the position of these "hot spots" where the electromagnetic field is locally amplified is the heart of intense research. In this perspective, we optimized the experimental conditions of our directed assembly setup to generate gold nanoparticles arrays as previously described but aiming at having the exact number of particles giving the maximum local electric field enhancement. We deposited a Raman label onto the substrate, and we performed some Raman measurements. We evidenced that, effectively, the electromagnetic field is locally amplified at the sites corresponding to the nanoparticle patterns. As we can observe in Fig. 7.10D,E, the intensity varies among the patterns, so the intensity of the enhanced Raman signal depends on several parameters, namely, the number of particles, their arrangement, their size, their orientation, etc. This substrate showed its great efficiency as a SERS but also as an MEF substrate. In fact, Fig. 7.10C shows that when a fluorophore is deposited at very low concentrations onto the substrate, fluorescence is locally amplified at the sites corresponding to the nanoparticles patterns only.

7.5 Perspectives for Patterning Virus Nanoparticles

During the last decade, a variety of highly organized nanoscale biological structures have inspired researchers to design new systems for producing novel nanomaterials. Among them, viruses are perfect examples of sophisticated supramolecular monodispersed assemblies with a remarkable plasticity in their structure and dynamics, including self-assembly and disassembly and site-specific delivery of cargo molecules. This knowledge has been translated into the idea of utilizing these natural virus nanoparticles (VNPs) in nanotechnology for various purposes. Because virus particles can serve as a molecular platform for assembling in a very controlled way other molecular entities, the question of their patterning becomes central. The capability of the technology to create controlled patterns of single VNPs will permit the design of advanced molecular nanodevices for sensing purposes for guiding the growth of highly organized novel nanomaterials to turn the specific properties of VNPs into practical uses.

One successful attempt to pattern viruses has been reported by Vega et al. [32]. In this work, tobacco mosaic virus (TMV) tubular particles were immobilized and positioned by combining dip pen nanolithography (DPN) and coordination chemistry, relying on the ability of metal ions to bridge a surface patterned with 16-thiohexadecanoic acid (MHA) molecules and the carboxylate-rich surface of TMVs. MHA was therefore patterned on gold surfaces using DPN, and the produced patterns served as a template to fix Zn^{2+} ions and TMV particles. Single-particle resolution could be achieved with this technology, but the processed surface could not be extend due to the relative low writing speed of the initial molecular patterns using the DPN approach. More recently, using self-organized nanopores formed inside an aluminium oxide membrane, Moon et al. [33] published a method for binding bacteriophage phi29 VNPs inside the produced nanopores with a reproducibility of 84%. Soft lithography was also proposed as a reliable method for producing large-scale arrays of single viruses by Solis et al. [34]. In this work by implementing a variant of the classical μCP method called subtractive printing, the authors showed that by patterning antibodies in feature sizes down to 100 × 100 nm, over a surface as large as 0.25 mm^2, and by controlling the concentration of M13 bacteriophages, they could produce large-scale arrays of sites equipped with single viruses with a success rate close to 100%. All these methods rely on the specific capture of the VNPs by molecular nanoscale patterns. Their efficiency and quality strongly depend on the quality of the surface chemistry and the nature of the VNPs under investigation. According to the results presented in this contribution, one may think that the combination of capillary assembly and soft lithography could be also a very efficient method to pattern well-controlled arrays of single VNPs in a versatile manner since it does not rely on any specific surface chemistry for their trapping. Indeed, by organizing single VNPs by capillary assembly in adapted topographical nanocavities at the surface of a PDMS stamp, it should be possible to transfer them by simple contact printing on any kind of surface. This method would conciliate low cost, high throughput, and versatility to any kind of VNPs independently of their molecular structure.

7.6 Conclusions

In this chapter, we have shown how soft lithography and more specifically μCP could be a crucial technique in the toolbox of nanotechnologies for biological applications. The combination of low-cost patterning at the nanoscale by μCP and label-free optical detection allowed us to propose a complete biochip platform. By reducing the cost of molecular analysis, we hope to penetrate the market of medical analysis as the proposed technology should give access to sophisticated molecular analysis at an affordable cost. In a more general view, soft lithography can be used to pattern nucleic acid molecules, lipids, proteins, nanotubes, nanoparticles, and various kinds of molecules largely used in surface chemistry, such as alkane-thiols, silanes, and poly-L-(lysine)-grafted-polyethylene(glycol), among others. These printing processes result in a sophisticated engineering of the surface that can be used to create, for example, patches of supported lipid bilayers [35, 36] for controlling cell adhesion and growth. More recently it has been also shown that μCP can even be employed to biofunctionalize preprocessed wafers by silicon technology such as nanoelectromechanical systems [37, 38].

We have also shown that soft lithography has become an invaluable technique at the service of single-object studies. This process is essential in the creation of chips with single cell, molecule, or particle resolution. We have set a versatile toolbox based on soft lithography combined with capillary assembly and incubation techniques to generate controlled arrays of single objects at the nanometer and the micrometer scale for different applications. We have shown that the strategies developed are very efficient to generate arrays of all kinds of objects ranging from cells and molecules to metallic nanoparticles with a high success rate in a precise, highly reproducible, large-scale, and cost-efficient manner. We have demonstrated that our methodology is capable of high throughput and is fully compatible with different types of supports, including glass or silicon. The assembled objects preserve their functionality and intrinsic properties and exhibit remarkable potential for a great number of major applications in the fields of medical analysis, gas sensing, and optics but also in the construction of tomorrow's nanoscale devices.

These few examples show that for over a decade, the emergence of nanobiotechnologies has motivated a new interdisciplinary scientific forum at the crossroads between molecular biology, physics, chemistry, and micro- and nanotechnologies, giving birth to some scientific breakthroughs and paving the way for industrial applications. The scientific context of this blooming field of research and innovation relies on three aspects: (i) the emergence of a technological offer for highly miniaturized, multifunctional integrated systems associating sensing devices, fluidics, and bioentities; (ii) the extraordinary progression in molecular biology, opening the route to huge scientific and application fields in genetics, epigenetics, proteomics, and medicine; and (iii) the development of a new physicochemistry of nano-objects. Indeed, molecular engineering through self-assembly has reached a level of maturity that is sufficient to enable a worldwide dissemination of functional nano-objects and nanostructured materials characterized by exquisite performances and broad application spectra.

Acknowledgments

This work is collective and highly interdisciplinary. Many collaborators, PhD students, and postdoctoral associates need to be deeply acknowledged. Special acknowledgments to Laurent Malaquin, Jean Christophe Cau, Childerick Séverac, Christel Martin-Cerclier, Mike Geneviève, Hélène Lalo, Jérôme Chalmeau, Adrian Martinez, Amélie Béduer, Amandine Egea, Angélique Coutable, Anna Labernadie, and Florent Seichepine for their incredible energy and talent in interdisciplinary research. Many colleagues of the University of Toulouse in chemistry, biology, and medicine are acknowledged for various projects built with them in the last 12 years. Their inputs were crucial for starting this activity, obtaining results, and projecting ourselves in the future.

References

1. Kumar, A., and Whitesides, G. M. (1993) Features of gold having micrometer to centimeter dimensions can be formed through a combination of stamping with an elastomeric stamp and an alkanethiol ink followed by chemical etching, *Appl. Phys. Lett.*, **63**, 2002–2004.
2. Biebuyck, H., Larsen, A., Delamarche, E., and Michel, B. (1997) Lithography beyond light: microcontact printing with monolayer resists, *IBM J. Res. Develop.*, **41**,159–170.
3. Gates, B., Xu, Q., Stewart, M., Ryan, D., Willson, C. G., and Whitesides, G. M. (2005) New approaches to nanofabrication: molding, printing, and other techniques, *Chem. Rev.*, **105**, 1171–1196.
4. Bernard, A., Fitzli, D., Sonderegger, P., Delamarche, E., Michel, B., Bosshard, H. R., and Biebuyck, H. (2001) Affinity capture of proteins from solution and their dissociation by contact printing, *Nat. Biotechnol.*, **19**, 866–869.

5. Renault, J. P., Bernard, A., Bietsch, A., Michel, B., Bosshard, H. R., Delamarche, E., Kreiter, M., Hecht, B., and Wild, U. P. (2003) Fabricating arrays of single protein molecules on glass using microcontact printing, *J. Phys. Chem. B*, **107**, 703–711.
6. Auletta, T., Dordi, B., Mulder, A., Sartori, A., Onclin, S., Bruinink, C. M., Péter, M., Nijhuis, C. A., Beijleveld, H., Schönherr, H., Vancso, G. J., Casnati, A., Ungaro, R., Ravoo, B. J., Huskens, J., and Reinhoudt, D. N. (2004) Writing patterns of molecules on molecular printboards, *Angew. Chem.*, **116**, 373–377.
7. Xia., Y., and Whitesides, G. M. (1998) Soft lithography, *Angew. Chem., Int. Ed. Engl.*, **37**, 550–575.
8. Thibault, C., Leberre, V., Casimirius, S., Trévisiol, E., François, J. M., and Vieu, C. (2005) Direct microcontact printing of oligonucleotides for biochip applications, *J. Nanobiotechnol.*, **3**, 7.
9. Thibault, C., Séverac, C., Mingotaud, A. F., Vieu, C., and Mauzac, M. (2007) Poly(dimethylsiloxane) contamination in microcontact printing and its influence on patterning oligonucleotides. *Langmuir*, **23**(21), 10706–10714.
10. Cau, J. C., Lalo, H., Severac, C., Peyrade, J. P., Trevisiol, E., Leberre, V., François, J. M., and Vieu, C. (2009) Molecular analysis for medicine: a new technological platform based on nanopatterning and label-free optical detection, *Oncologie*, **11**, 1–5.
11. Egea, A., Mazenq, L., Trévisiol, E., Paveau, V., and Vieu, C. (2013) Optical label free biodetection based on the diffraction of light by nanoscale protein gratings, *Microelectron. Eng.*, **111**, 425–427,
12. Cerf, A., Cau, J. C., and Vieu, C. (2008) Controlled assembly of bacteria on chemical patterns using soft lithography, *Coll. Surf. B*, **65(2)**, 285–291.
13. Cerf, A., Cau, J. C., Vieu, C., and Dague, E. (2009), Nanomechanical properties of dead or alive single-patterned bacteria, *Langmuir*, **25(10)**, 5731–5736.
14. Cerf, A., Thibault, C., Geneviève, M., and Vieu, C. (2009) Ordered arrays of single DNA molecules by a combination of capillary assembly, molecular combing and soft lithography, *Microelectron. Eng.*, **86**(4–6), 1419–1423.
15. Cerf, A., and Vieu, C. (2009) Transfer printing of sub-100 nm nanoparticles by soft lithography with solvent mediation, *Coll. Surf. A*, **342**(1–3), 136–140.
16. Lange, S., Benes, V., Kern, D., Horber, H., and Bernard, A. (2004) Microcontact printing of DNA molecules, *Anal. Chem.*, **76**, 1641–1647.

17. Ferrari, M. (2005) Cancer nanotechnology: opportunities and challenges, *Nat. Rev.*, **5**, 161–171.

18. Thibault, C., Severac, C., Trévisiol, E., and Vieu, C. (2006) Microtransfer moulding of hydrophobic dendrimer, *Microelectron. Eng.*, **83**, 1513–1516.

19. Lalo, H., and Vieu, C. (2009) Nanoscale patterns of dendrimers obtained by soft lithography using elastomeric stamps spontaneously structured by plasma treatment, *Langmuir*, **25**(13), 7752–7758.

20. Vieu, C., Carcenac, F., Pépin, A., Chen, Y., Mejias, M., Lebib, A., Manin-Ferlazzo, L., Couraud, L., and Launois, H. (2000) Electron beam lithography: resolution limits and applications, *Appl. Surf. Sci.*, **164**, 111.

21. Chalmeau, J., Thibault, C., and Vieu, C. (2008) Self-aligned patterns of multiple biomolecules printed in one step, *Appl. Phys. Lett.*, **93**, 133901–133903.

22. Chalmeau, J., Thibault, C., Carcenac, F., and Vieu, C. (2008) Micro-contact printing of two different biomolecules in one step using deformable poly(dimethylsiloxane)-based stampjapanese, *J. Appl. Phys.*, **47**, 5221–5225.

23. Lalo, H., Cau, J. C., Thibault, C., Marsaud, N., Severac, C., and Vieu, C. (2009) Microscale multiple biomolecules printing in one step using a PDMS macrostamp, *Microelectron. Eng.*, **86**, 1428–1430.

24. Akou, A., Gural'skiy, I. A., Salmon, L., Bartual-Murgui, C., Thibault, C., Vieu, C., Molnár, G., and. Bousseksou, A, (2012) Soft lithographic patterning of spin crossover complexes. Part 2: stimuli-responsive diffraction grating properties, *J. Mater. Chem.*, **22**(9), 3752–3757.

25. Cau, J. C., Cerf, A., Thibault, C., Séverac, C., Peyrade, J. P., and Vieu, C. (2008) Chemical engineering of silicon oxide surfaces using micro-contact printing for localizing adsorption events of nanoparticles, dendrimers and bacteria, *Microelectron. Eng.*, **85**(5–6), 1143–1146.

26. Labernadie, A., Thibault, C., Vieu, C., Maridonneau-Parini, I., and Charrière, G. (2011) Dynamics of podosome stiffness revealed by atomic force microscopy, *PNAS*, **107**(49), 21016–21021.

27. Labrousse, A. M., Meunier, E., Record, J., Labernadie, A., Beduer, A., Vieu, C., Ben Safta, T., and Maridonneau-Parini, I. (2011) Frustrated phagocytosis on micro-patterned immune complexes to characterize lysosome movements in live macrophages, *Front. Immun.*, **2**, 51.

28. Béduer, A., Seichepine, F., Flahaut, E., and Vieu, C. (2012) A simple and versatile micro contact printing method for generating carbon

nanotubes patterns on various substrates, *Microelectron. Eng.*, **97**, 301–305.

29. Béduer, A., Seichepine, F., Flahaut, E., Loubinoux, I., Vaysse, L., and Vieu, C. (2012) Elucidation of the role of carbon nanotube patterns on the development of cultured neuronal cells, *Langmuir*, **28**(50), 17363–17371.
30. Cerf, A., Molnar, G., and Vieu, C. (2009) Novel approach for the assembly of highly efficient SERS substrates, *ACS Appl. Mater. Interfaces*, **1**(11), 2544–2550.
31. Nie, S. M. & Emory, S. R. (1997) Probing single molecules and single nanoparticles by surface-enhanced raman scattering, *Science*, **275**, 1102–1111.
32. Vega, R. A., Maspoch, D., Salaita, K., and Mirkin, C., (2005) Nanoarrays of single virus particles, *Angew. Chem., Int. Ed.*, **44**, 6013–6015.
33. Moon, J. M., Akin, D., Xuan, Y., Ye, P. D., Guo, P., and Bashir R. (2009) Capture and alignment of phi29 viral particles in sub-40 nanometre porous alumina membranes, *Biomed. Microdevices*, **11**, 135–142.
34. Solis, D. J., Coyer, S. R., Garcia, A. J., and Delamarche, E. (2010) Large-scale arrays of aligned single viruses, *Adv. Mater.*, **22**, 11–114.
35. Chalmeau, J., Salomé, L., Thibault, C., Severac, C., and Vieu, C. (2007) Elaboration of micro domains of supported bilayer membrane using microcontact printing, *Microelectron. Eng.*, **84**, 1754.
36. Chalmeau, J., LeGrimellec, C., Sternick, J., and Vieu, C.(2012) Patterned domains of supported phospholipid bilayer using microcontact printing of Pll-g-PEG molecules, *Coll. Surf. B*, **89**, 188–195.
37. Guillon, S., Salomon, S., Seichepine, F., Dezest, D., Mathieu, F., Bouchier, A., Mazenq, L., Thibault C., Vieu, C., and Nicu, L. (2012) Biological functionalization of massively parallel arrays of nanocantilevers using microcontact printing, *Sens. Actuators, B*, **161**(1), 1135–1138.
38. Salomon, S., Leichlé, T., Dezest, D., Seichepine, F., Guillon, S., Thibault, C., Vieu, C and Nicu, L. (2012) Arrays of nanoelectromechanical biosensors functionalized by microcontact printing, *Nanotechnology*, **23**, 495–501.

Index

activity 47, 58, 68, 72, 78, 79, 81, 90, 115, 124, 136, 163, 164, 169–171, 176, 202, 233
- anticoagulant 202
- biocatalytic 141
- biological 53, 83
- catalase 171
- cell cycle 81
- enzymatic 11, 69, 140, 141, 173, 200, 202

AFM *see* atomic force microscopy
amino acids 7, 50, 53, 62, 66, 73, 75, 80–82, 114, 126, 173
- artificial 61, 63
- canonical 62
- natural 50, 59
- reactive 54

antibodies 23, 27, 72, 90, 91, 171–173, 202, 217, 222
- bifunctional 89
- monoclonal 89, 160, 162
- single-chain 80

antigens 26, 90, 176, 198
- human leukocyte 197

aptamers 106, 193, 197, 198, 202
assembly 5, 7, 8, 10, 11, 13, 15–17, 20, 66, 69, 71, 73, 106, 113, 114, 118, 200, 201, 225–228
- biosynthetic 106
- capillary 208, 231, 232
- polymeric 130
- supramolecular 53, 66, 106
- symmetrical 107
- virion 5, 12, 13
- virus protein–cage 106

assembly process 5–7, 16, 223, 228
atomic force microscopy (AFM) 21, 199, 222, 225

bacteria 25, 48, 62–64, 68, 69, 88, 89, 224
bacterial viruses 29, 48
bacteriophages 3, 8, 20, 21, 25, 48, 53, 63, 67, 69, 88, 110, 113, 115, 116, 231
- filamentous 27
- head–tail 24
- tailed 10

BBMV *see* broad bean mottle virus
biomolecules 85, 202, 208–210, 213, 218, 219
blue tongue virus (BTV) 21, 25
BMV *see* brome mosaic virus
broad bean mottle virus (BBMV) 17
brome mosaic virus (BMV) 8, 13, 14, 17, 22
BTV *see* blue tongue virus

CAMs *see* cellular adhesion molecules
capsid 1–3, 6, 12–14, 21, 26, 53, 60, 68, 73, 74, 106, 109–111, 113, 115–118, 147, 148
- adenoviral 23, 24, 71, 72
- four-helix-bundle 29
- mature 110
- native 115
- viral 14, 71, 90
- wild-type P22 25

wild-type T7 68
capsid protein 15, 50, 106, 113, 114
cargos 71, 77, 134, 164, 165, 167, 169, 171, 173, 177, 179, 198
benign 172
cellular 14
drug 161
enzymatic 181
fluorescent 71
therapeutic 160
carriers 163, 169, 171, 180, 181
lipid-based 163
lipophilic ion 140
micron-sized 178
polymeric 171
submicron 178
traditional 171
cascade reactions, multiple-step 123
catalysts 59–61, 110, 125, 127, 129, 136, 139, 145, 148, 149
CCMV *see* cowpea chlorotic mottle virus
CCMV capsids 22, 109, 116
cell-internalizing peptide (CIP) 86
cells 12–14, 62, 70, 71, 76–79, 82, 85, 105, 116–119, 160, 169, 171, 172, 178, 179, 201, 202, 208, 223, 224
artificial 127, 128, 134, 149
bacterial 224
blood 167
hepatic 85
human 225
immune 160
infected 79
insect 83, 85
intestinal 73
natural 138
transfected 77
yeast 85
cellular adhesion molecules (CAMs) 160
CIP *see* cell-internalizing peptide
clots 164–167
canine 166
human blood 166
occlusive 167
CLPs *see* core-like particles
coat proteins (CP) 1, 7–9, 11–14, 16, 26, 27, 46, 50, 55, 62, 63, 66, 67, 69, 88, 110
concentrations 79, 106, 111, 116, 117, 134, 177, 214–216, 225, 231
critical micelle 130
glucose 133, 134
plasma urate 174
proton 13
contact angle 211, 212
core-like particles (CLPs) 46, 73, 75–77
coupling 47, 50, 53, 59, 62, 64, 200, 229
cowpea chlorotic mottle virus (CCMV) 13, 17, 22, 48, 63, 89, 107–109, 111, 115–118, 147, 148
CPs *see* coat proteins
CPs
potyviral 19
wild-type 66
crosslinking 55, 56, 129, 139, 211
biocompatible 59
disulfide 52
subunit 47
tyrosine–tyrosine 55

degradation 19, 60, 165, 167, 173, 178, 179
biofilm 69

glycogen 179
proteolytic 160, 162, 164, 171
dimers 21, 22, 109, 201
dip pen nanolithography (DPN) 230
DKR *see* dynamic kinetic resolution
DNA molecules 208, 211, 213, 214, 216, 217, 225–227
DNA origamis 193, 199–202, 209
double-stranded DNA 10, 89, 134, 199
DPN *see* dip pen nanolithography
drugs 85, 138, 161, 163, 170, 175, 178
efficacious 159
enzymatic 180
labile 140
small-molecule 159
dynamic kinetic resolution (DKR) 127, 132, 133

encapsulation 25, 114–117, 132, 138, 140, 145, 170, 171, 174
ENCs *see* enzyme nanocarriers
endoplasmic reticulum (ER) 76, 82, 85
endothelial cells 160, 168, 171, 172, 178
enveloped viruses 3, 28, 64, 75, 76, 85
envelope proteins 28, 46, 79, 80, 85
environments 3, 5, 29, 129, 147, 209
anaerobic 60
confined 105–107, 114, 115, 117–119
proton-enriched 144
research laboratory 216
stable 141
enzymatic reactions 124, 141, 172
enzyme activity 47, 78, 118, 177
enzyme cargo 111, 113
enzyme classes 177, 180
enzyme efficiency 115, 116, 118
enzyme loading 111, 113, 114, 163
enzyme nanocarriers (ENCs) 5, 25, 27, 48, 66
enzyme replacement therapy (ERT) 160, 175, 177, 180
enzymes 26, 27, 45–51, 55, 57–64, 68–72, 77–79, 88–90, 105–107, 113–119, 126–128, 139–144, 159–162, 164–166, 169–178, 200
citotoxic 81
conjugated 53
detoxifying 160, 174
encapsulated 113, 115, 116, 173
free 115–117, 171, 173, 179, 202
lipase 27
mamalian 86
proteolytic 142, 164, 171
tumor inhibitory 172
ER *see* endoplasmic reticulum
ERT *see* enzyme replacement therapy

fluorophores 68, 71, 217, 230
commercial 213
light-emitting 217
fusion 24, 25, 46, 67, 72, 78, 79, 85, 87, 111
coat protein–enzyme genetic 64
enzyme–CP 27
lactamase 68

peptide–CP genetic 66
fusion proteins 66, 72, 79, 81, 87, 90, 160

genes 70, 77, 79, 83, 85, 86, 129, 137, 212, 217
biosynthetic 126
immune 20
pol 78
therapeutic 72
yeast 212
genetic material 64, 70, 76, 106, 147
genomes 2, 3, 5, 6, 8, 10, 13, 14, 22, 26, 29, 79
bipartite 15
infectious 11
monopartite 15
nucleic acid 1
viral progeny 14
GFP *see* green fluorescent protein
gold nanoparticles 135, 193, 197, 198, 211, 228, 229
green fluorescent protein (GFP) 67, 70, 72, 73, 75, 78, 80, 137
groups
alkyne 60
carboxyl 17
ester 49
heparin 167
nitrene 49
phosphatidylinositol 82
photoreactive phenyl azide 46, 49
thiol 52
vinyl 170

HBV *see* hepatitis B virus
hepatitis B virus (HBV) 21, 63, 73, 74, 85
HIV *see* human immunodeficiency virus
HLA *see* human leukocyte antigen
host cells 1, 29, 75, 83, 114
human immunodeficiency virus (HIV) 8, 21, 80
human leukocyte antigen (HLA) 176, 197

infection 11, 14, 19, 76, 114
injection
heparin 168
intramuscular 175
intraperitoneal 177
intravenous 171
interactions 8, 11, 14, 75, 79, 87, 90, 91, 124, 130, 132, 161, 164, 217, 220, 228
antibody–antigen 200
biomolecular 220
carboxyl–carboxylate 13
charge–charge 111
dimer 21
electronic 132
electrostatic 132
hydrophobic 16
hydrophobic–hydrophilic 137
molecular 6, 21, 209
phosphate–carboxylate 13
probe–target 221
protein–antibody 220
protein–nucleic acid 11
protein–protein 6, 17, 21, 106
protein–RNA 17, 22
steric 202
virus–host cell 14

layers 132, 133, 139
aqueous 128
charged 132
concentric 76
core capsid 2

dendrimer 215
hydration 200
lipid bilayer 2, 28, 75, 137, 162
lipid rafts (LRs) 81–83
lipids 11, 178, 232
liposomes 137–139, 149, 162, 165, 166, 169, 170, 173, 177, 178, 180
antioxidant 169
conventional 162
multicompartment 137
phospholipid 139
living cells 123, 128, 129, 134, 137, 138, 141, 148, 149, 201, 223, 224
LRs *see* lipid rafts
LSDs *see* lysosomal storage disorders
lysine 48, 53, 54, 86, 87, 232
lysine residues 48, 50, 52, 53, 55, 61, 87, 164
lysosomal storage disorders (LSDs) 159, 177, 178, 180

magnetic nanoparticles (MNPs) 162, 163, 167, 171
major capsid proteins (MCPs) 3, 15, 22, 23, 25, 26
MCPs *see* major capsid proteins
membrane 5, 46, 82, 105, 127, 129, 137, 139–143, 146, 149
aluminium oxide 231
cellular 28, 162
cytoplasmic 75, 77
liposomal 137
permeable 129, 132, 140, 141
plasma 76, 81–83, 87
polymer 141
semipermeable 134
vesicle 29, 139
membrane localization sequences (MLSs) 85
metal catalysts 127–130, 143
microdroplets 128, 129
MLSs *see* membrane localization sequences
MNPs *see* magnetic nanoparticles
model 6, 24, 90, 91, 107, 118, 170
animal 169, 180
canine 165
ex vivo 168
molecular 91
mouse arterial thrombosis 168
paraquat-induced oxidative injury 170
rabbit jugular vein thrombosis 165
rat 166
rat adjuvant rat arthritis 170
rat arteriovenous shunt thrombosis 167
rat embolic 167
rat jugular vein thrombosis 167
molecules 47, 123, 132, 141, 147, 194, 200, 202, 210, 211, 215, 218, 219, 224, 227, 229, 230, 232
active 73
cargo 230
complex 123
contaminating 213
docking 224
growth-stimulating 172
hydrophobic 223
inhibitory 110
ink 218

nanoparticles 20, 73, 86, 106, 163, 167, 168, 173, 176, 180, 216, 228, 232

chitosan glutamate 176
heterogeneous 135, 139
hybrid 66
metallic 232
polymer 129
silica 139
nanoreactors 23, 71, 107, 109, 110, 129, 132, 142, 143, 147, 149, 176, 177
multicompartment 134
viral 147
yolk-shell 136
NCs *see* nucleocapsids
NCs
anti-ICAM 179
anti-PECAM-coated 171
conventional 161
enzyme-loaded 178
fibrinolytic 164
flow-sensitive 169
nonspherical 164
Niemann Pick disease (NPD) 178, 179
NLS *see* nuclear localization signal
NPC *see* nuclear pore complex
NPD *see* Niemann Pick disease
nuclear localization signal (NLS) 13, 78
nuclear pore complex (NPC) 13, 14
nucleic acids 68, 76, 85, 163, 193, 199, 202
nucleocapsids (NCs) 5, 11, 76, 159–161, 163–166, 169, 171, 172, 174, 175, 178–180

oligonucleotides 202, 211–213, 215
organelles 105, 118, 119
oxidation 56, 60
enzymatic 134, 142
fatty acid 123

particles 4, 5, 13, 14, 17, 18, 20–22, 24, 25, 48, 61, 73, 74, 77–79, 84–86, 134–136, 162, 163, 166–168, 176–178, 228–230
adenovirus 23
baculovirus 20
charged 160
chimera 76
colloidal 132, 146
disk-shaped 178
double-layered 73
double-shelled 17
ellipsoidal 17
enveloped 76
enzyme-carrying 23
filamentous 27
hollow 85
icosahedral 53
infectious 80
nonenveloped 46
prolidase-loaded 176
rod-shaped 16
swollen 17
triple-layered 73
tubular 230
viral 47
yolk-shell 136
PCR *see* polymerase chain reaction
PDMS stamp *see* poly(dimethylsiloxane) stamp
PDMS stamp
cleaned 215, 216
flat 211–213
hydrophobic 211
microstructured 226, 227
multilevel 219
nanostructured 228

structured 225, 227
uncleaned 214
PECAMs *see* platelet endothelial cell adhesion molecules
PEGylation 164, 165, 180
peptides 4, 23, 26, 27, 54, 66–68, 71, 74, 86, 88–90, 160, 162
adaptor 87
arginine-rich 88
cell-penetrating 70
coiled coil 111
conjugate 24
fusogenic 88
helical linker 66
immunogenic 72
linear 75
nanoglue 86
streptavidin fusion 26
plants 14, 19, 26, 53, 55, 107, 110
infected 53
viruses infecting 25
wild 29
plant viruses 12, 13, 15, 26, 29, 63, 66, 76
platelet endothelial cell adhesion molecules (PECAMs) 171, 172
poly(dimethylsiloxane) stamp (PDMS stamp) 212–215, 218, 219, 223, 225, 226, 228, 231
polymerase chain reaction (PCR) 134
polymers 130, 139, 141, 145, 163
amphiphilic 141, 145
biological 163
cholesterol-based 139
crosslinkable 139
hydrophilic 139, 162
linear 130
polymersomes 106, 139–147, 149
post-translational modifications (PTMs) 19, 76, 86
potato virus X (PVX) 67
potato virus Y (PVY) 66
procapsids 10, 11, 71, 110
proline-rich region (PRR) 80
properties 3, 4, 20–23, 25, 27, 55, 64, 129, 139, 161, 193, 197, 228–230
biological 5
bonding 6
elastomeric 219
intrinsic 228, 232
mechanical 199
membrane fusion 28
physical 4
proteins 5, 23, 45, 46, 49, 50, 55, 59, 61–66, 69–71, 75, 77–87, 89–91, 148, 163–165, 193, 200–202
active 144
bacterial channel 140, 176
cargo 71, 114
chimeric 86
complex 20, 28, 76
fiber knob 23, 71
glue 5
glycosylated 19
heterelogues 78
influenza virus M2 86
monomeric 84
monomeric spike 76
neck 69
nucleocapsid 90
pore-forming 137
soluble 19
spike 90
tail 69
therapeutic 162
vehiculate bioactive 78
viral membrane 28
viral replicase 12
virus-encoded 6, 15
protein subunits 2, 4, 7–10, 15, 20, 147

PRR *see* proline-rich region
PTMs *see* post-translational modifications
PVX *see* potato virus X
PVY *see* potato virus Y

reactions 49, 52, 57, 59–61, 87, 123–127, 129–131, 139, 140, 143, 149, 169, 174
  biochemical 105
  catalytic 140, 148
  cycloisomerization 131
  immune 164, 165
  inflammatory 177
  organic 130
reactive oxygen species (ROS) 169, 172
reagents 47, 87, 124, 125, 127–129, 131, 140, 149
  bifunctional 49
  incompatible 127
  nonharmful 127
  sensitive 129
  thiolation 52
receptors 5, 28, 68, 70, 75, 78, 82
  cellular 23
  cognate 80
  low-affinity nerve growth factor 87
  natural 80
  tumor 68
RNA 6–10, 13, 14, 17, 18, 22, 106, 134, 199
  double-stranded 10
  genomic 13, 113
  single-stranded 7
ROS *see* reactive oxygen species

SAMs *see* self-assembled monolayers
SANS *see* small-angle neutron scattering
scaffold protein (SP) 25, 71, 110, 113
scaffolds 47, 90, 107, 110, 194, 195, 197, 198
self-assembled monolayers (SAMs) 210
signals 9, 78, 86, 90, 199, 214, 216
  environmental 2
  nuclear 78
  nuclear localization 13, 78
  Raman spectroscopic 229
simian immunodeficiency virus 81, 83
sites 47, 48, 62, 64, 71, 82, 124, 160, 162, 167, 169, 175, 180, 230, 231
  acceptor 75
  catalytic 130, 149
  cleveage 78
  protease 78
  reactive 138
  receptor-binding 80
  ribosome-binding 88
  subcellular 160
  thrombin-cleavage 71
  viral protease 81
SIV *see* simian immunodeficiency virus
small-angle neutron scattering (SANS) 21
soft lithography 207, 208, 210, 218, 223, 225, 231, 232
SP *see* scaffold protein
species 46, 55, 176, 223
  biological 209
  fluorescent 134, 143
  homopolymers 46
  reactive 46
  reactive oxygen 169, 172
  viroid 3
SPRs *see* surface plasmon resonances
strategies 26, 27, 45, 46, 48, 51, 55, 59, 60, 64–67, 84, 88, 89, 91, 124, 126, 129, 172, 174

origami 193, 194
split-core 73, 74
two-pronged 69
versatile 75
structure
actin-rich 225
barrel 22
biological 12, 55, 230
disk 10
lipid 83
liposome 170
macromolecular 15
membranous 12
molecular 48, 231
planar 194, 200
supercoiled 106
tail 15
topographical 219
substrates 23, 105, 107, 109, 110, 114, 115, 117, 118, 124, 127, 129, 135, 140, 147–149, 176, 177, 218, 228–230
fluorescent 68, 86
glass 213
hydrophilic 211
tyrosinyl 55
surface plasmon resonances (SPRs) 229
systems 19, 20, 25, 73, 75, 78, 79, 114, 119, 139, 145, 160, 207–209
artificial 209
bacterial expression 19
baculovirus expression 20
baculovirus–insect cell 73
biphasic 128
catalytic 60, 124
complex 138
heterologous 48, 73
heterologous expression 25
homobifonctional 50
immune 175
liposomal 162
mammalian 19, 20
nanoelectromechanical 232
reticuloendothelial 160
ribosomal 88
silica-coated 136
viral expression 19
yeast 19

TCV *see* turnip crinkle virus
techniques 20, 21, 27, 127, 128, 132, 141, 199, 210, 218, 223, 225, 228, 229
fabrication 132
incubation 223
macroscopic separation 127
optical 229
TEM *see* transmission electron microscopy
thrombosis 159, 165, 178
TMV *see* tobacco mosaic virus
tobacco mosaic virus (TMV) 2, 6, 8, 13, 15, 26, 48, 107, 108, 230
transmission electron microscopy (TEM) 21, 90
turnip crinkle virus (TCV) 6
turnip vein clearing virus (TVCV) 66
TVCV *see* turnip vein clearing virus

UAAs *see* unnatural amino acids
unnatural amino acids (UAAs) 61, 62, 64, 65

vesicles 12, 105, 106, 131, 132, 134, 135, 137, 139–141, 144, 149, 161
aqueous 147
lipid 29, 137
lipid-based 168
liposome-in-polymersome 138

micellar 131
multicompartment 138, 146, 149
phospholipid bilayer 162
polymeric 144
robust 139
shell-in-shell 149
single-compartment 138
yolk-shell 135
vesicular stomatitis virus (VSV) 28, 79
viral genomes 1, 4, 5, 66, 90
including 5
viral nanoparticles (VNPs) 3–5, 45–47, 50, 53, 58, 66–68, 230, 231
viral painting 83, 84
viral particles (VPs) 47, 48, 53, 55, 57, 58, 64, 80, 81, 84
viral proteins 24, 28, 76
viral RNA 7, 8, 11, 19, 88
virions 2–7, 11, 13, 15, 17, 18, 20–22, 81, 88
flexuous 15
infectious 11, 13, 15
mature 6
rod-shaped 26
virus 3, 5, 8, 16, 17, 20, 21, 28, 29, 47, 48, 53, 68, 69, 73, 76, 106–108, 110, 111, 113–115, 117, 118
avian leucosis 28
avian sarcoma leukosis 79
biotinylated 87
blue tongue 25
cardamom mosaic 19
cauliflower mosaic 13
coil-shaped 29
feline herpes 84
foot-and-mouth disease 66
human immunodeficiency 8, 80
icosahedra 17
infectious 10
mouth disease 27
murine leukemia 28
pepino mosaic 67
recombinant 48
simian 113
spindle-shaped 29
stable 110
tomato bushy stunt 3
two-tailed 29
vesicular stomatitis 28, 79
virus assemblies 6, 109, 111, 113, 116
virus capsids 1, 106, 107, 113, 114, 117
virus genome 1, 48, 88
virus-like particles (VLPs) 4, 15, 17–21, 63, 64, 69–71, 77–79, 81–85, 88, 89, 106, 147, 148
virus particles 1–4, 6, 15, 23, 28, 91, 147, 230
filamentous 27
icosahedral 5
progeny 1
virus species 2, 21
VLPs *see* virus-like particles
VLPs
alphaviruses-derived 77
empty 89
mature 71
stable 73
VNPs *see* viral nanoparticles
VNPs
hybrid 67
infectious 4
purified 48
spherical 66
VPs *see* viral particles
VPs
icosahedral 63
purified 48
VSV *see* vesicular stomatitis virus

An environmentally friendly book printed and bound in England by www.printondemand-worldwide.com

PEFC Certified

This product is from sustainably managed forests and controlled sources

www.pefc.org

PEFC

PEFC/16-33-415

This book is made of chain-of-custody materials; FSC materials for the cover and PEFC materials for the text pages.